Seyyed Hossein Nasr's Ecological Ethics

Bridging Science, Religion, and the Environment

By

Muzzamel Hussain Imran

Seyyed Hossein Nasr's Ecological Ethics: Bridging Science, Religion, and the Environment

By Muzzamel Hussain Imran

This book first published 2023

Ethics International Press Ltd, UK

British Library Cataloguing in Publication Data

A catalogue record for this book is available from the British Library

Print Book ISBN: 978-1-80441-294-7

eBook ISBN: 978-1-80441-295-4

Table of Contents

Introduction

Seyyed Hossein Nasr, an Iranian-American scholar of Islam and comparative religion, born in 1933, is a prominent figure in the field of Islamic views on ecological matters. He is widely recognized as a leading authority and possibly the "founding father" of Islamic environmentalism. Nasr's work has paved the way for ongoing debates and discussions on the relationship between Islam and the environment.[1] He was among the first to approach the topic from an Islamic perspective and spent a significant part of his life in the US, where he raised awareness of the spiritual aspects of the ecological crisis in the Western world. In 1965, several months before Lynn White Jr's famous book on the "historical roots of our ecological crisis," Nasr published an essay on the subject.[2]

Despite Nasr's persistent efforts in writing and speaking about the topic, his message has not garnered the same level of attention as Lynn White's book. White blamed Christianity for the ecological crisis and played a significant role in shifting Western cultural views away from religion. On the other hand, Nasr's views advocating a return to genuine religion may have been less well-received. For reasons unknown, Nasr's ecological perspectives seem to have gone largely unnoticed until now. Nasr's reputation is extensive and acknowledged in the field of Islamic philosophy, comparative religion, and the interplay between Islam and modernity. He is also celebrated for his holistic perspective on knowledge, science, and the sacred, making him a commonly cited source for both Muslim and non-Muslim individuals. Nasr's written works on Islamic cosmology, mysticism, and metaphysics, have established him as a respected contemporary Iranian philosopher. His contribution to academia has earned him high regard within academic circles.[3] Furthermore, Nasr has emerged as a distinguished personality behind the contemporary program "*A Common*

---

1 B. Taylor (ed.), "Encyclopedia of Religion and Nature" (New York: Continuum, 2005-2008), s.v. "Seyyed Hossein Nasr."

2 In this book, the "ethos of medieval Christianity" is linked to the emergence of "an exploitative attitude towards nature in the Western World." For further information, see: B. Taylor (ed.), "Encyclopedia of Religion and Nature" (New York: Continuum, 2005-2008), s.v. "White, Lynn"; Lynn White, "The Historical Roots of Our Ecological Crisis," Science 155:3767 (March 10, 1967): 1203-1207.

3 M. Fakhry, "A History of Islamic Philosophy" (New York: Columbia University Press, 2004), 322.

Word," aiming to cultivate optimistic communication with the Catholic Church. Various publications have recently surfaced, scrutinizing and evaluating the perception of Nasr's scientific contributions in Indonesia, along with his stance on religion, diversity, interfaith conversation, and association with Traditionalism.[4] [5] Around the turn of the 21st century, a cohort of academics took notice of Nasr's ecological teachings and initiated a study of the interplay between humanity, its surroundings, and the spiritual realm. Nasr's ecological contributions have subsequently been recognized in the authoritative *Encyclopedia of Religion and Nature.* [6] Despite the valuable perspectives these sources provide, a holistic evaluation of Nasr's ecological viewpoints remains absent.

The objective of this book is to analyze and contextualize Nasr's ecological thought, particularly focusing on the crucial role played by the concepts of "religion," "tradition," and "modernity" in his argument. According to Nasr, to solve the ecological crisis, the contemporary world must reexamine traditional principles and knowledge concerning nature and the cosmos. He believes that only "traditional cosmologies" hold the key to solving this crisis, which can be found at the core of all genuine religions or traditions, but must be retrieved from their often neglected heritage. In this book, we will investigate how Nasr's understanding of traditional cosmologies can be situated within the frameworks of "traditionalism" or "perennial philosophy." This inquiry will underscore the belief in the "intrinsic harmony of all religions," which is closely linked to intellectual movements in the West, including "Western esotericism." A precise definition is required from an academic perspective in order to tackle these challenging categories.

Nasr asserts that Islam holds a critical significance in the present global context, as the onset of modernity has led to a significant loss of traditional knowledge in Western society, while Islamic communities maintain a strong connection to the sacred. Therefore, Nasr believes that Islam offers valuable

4 "A Common Word", accessed March 25, 2023, www.acommonword.com/.

5 Adnan Aslan, "Religious Pluralism in Christian and Islamic Philosophy, The Thought of John Hick and Seyyed Hossein Nasr" (Richmond: Curzon Press, 1998). Seyyed Hossein Nasr on "Science and the Reception of his Ideas in Indonesia," MA book, Leiden University, 2005,

6 Bron Taylor (ed.), "Encyclopedia of Religion and Nature, 2 vols." (New York: Continuum, 2005, 2008), s.v. "Nasr, S.H.," "Perennial philosophy," "Islam," "Islam and eco-justice," "Islam and environmental ethics."

insights that could benefit the West. This research will primarily focus on Nasr's perspective on the "Islamic tradition," the Islamic approach to environmental issues, and the ethical principles of Islam regarding the environment. According to Nasr, traditional Islam encompasses both an esoteric and exoteric aspect, with inner and outer dimensions. The non-Western manifestation of Western esotericism can be observed in the neo-platonic, mystical, gnostic, or theosophical philosophy that characterizes the inner dimension of Sufism. In contrast, the outer aspect follows orthodox practices and involves adherence to traditional *shari'a* laws concerning religious rituals (*ibadat*) and social interactions (*mu'amalat*), rooted in the Quran and Hadith. However, Nasr's concept of Islam differs from that of fundamentalists, as it encompasses not only these laws but also the cultural accomplishments of fourteen centuries of Islamic civilization, including philosophy, arts, poetry, architecture, and urban planning.

Nasr's Islamic environmental ethics are rooted in the beliefs mentioned above. Although his views are comprehensive and do not fully address practical implementation, they serve as a guide to the emerging fields of Islamic environmental law, economics, and Islamic reformist political theories. These fields represent the most direct application of Nasr's Islamic perspective and can be viewed as practical interpretations of his metaphysical cosmology. The relationship between worldview and practice in the Islamic discourse on the environment is significant. Nasr, among many others, posits that Islam provides a comprehensive worldview and the means to translate cosmology into practice, which he describes as connecting the Heavens to the Earth. These means, derived from the classical sharia (madhab), facilitate the establishment of ethical norms for environmental concerns. However, this gives rise to a politically charged environment with divergent and conflicting opinions on the application of *Sharia* in the present-day context. This book aims to contextualize Nasr's perspectives within the wider range of current Islamic ideologies.

Contemporary Ecological Crisis

According to substantial studies, the causes of recent global warming may mostly be attributed to anthropogenic sources or human actions that have a significant negative influence on the environment.[7] There are also grave concerns about the effects on human society as a whole. These human actions are regarded as environmentally important behaviors. Environmentally relevant behaviors can be defined by their effect, which refers to how much they affect resource availability and the composition of the biosphere, or by their purpose, which refers to action conducted with the intention of changing the environment.[8] The process of modernity has caused humanity to turn outward, making it difficult for them to comprehend that the environmental destruction they are causing is a manifestation of the internal destitution of their own souls.[9] Stott discusses the consequences of pollution on humanity, arguing that it poses a greater threat than even nuclear war. He identifies four major factors that are contributing to the devastation of the natural environment: population growth, depletion of resources and biodiversity, improper waste disposal, and the release of toxic chemicals into the atmosphere, which affects the protective ozone layer. According to Stott, these factors are interconnected and constitute a singular "interlocking global crisis."[10]

Since the advent of large-scale agriculture and human habitation, the world's forests have suffered immense devastation. Unfortunately, the damage has continued to the present day, and if deforestation trends persist, it is projected that nearly 25% of all species will become extinct by 2025. In addition to their intrinsic value, forests play a vital role in supporting human life by providing food, fiber, medicine, industrial goods, and genetic resources necessary for breeding better crop varieties, which in turn, support global food security.[11] However, human activities have led to rising levels of carbon dioxide, methane, and CFCs in the atmosphere, contributing to climate change. This, in turn, is causing the polar ice caps to melt, which will lead to a rise in sea levels, disrupting agriculture and

---

7 Paul C. Stern, "New Environmental Theories: Toward A Coherent Theory Of Environmentally Significant Behavior," Journal Of Social Issues 56, no. 3 (2000): 407-424, doi:10.1111/0022-4537.00175.

8 Ibid.

9 Seyyed Hossein Nasr, "Man and Nature: The Spiritual Crisis in Modern Man" (Chicago: ABC International Group, 1997), 3.

10 John R. W Stott, "Wakai No Tsutome" (repr., Tōkyō: Sugu Shobō, 2001).

11 Jessica Tuchman Mathews, "Preserving The Global Environment" (repr., New York: W.W. Norton, 1991), 1.

shifting ecological habitats across the landscape. Unless effective measures are taken to mitigate these impacts, many species that are unable to adjust to the changing climate and its effects could face extinction.[12]

The environmental and ecological crisis of today is largely attributed by climate scientists to the economies and lifestyles based on contemporary science and technology.[13] The crisis comprises various issues such as air pollution caused by burning fossil fuels (oil, gas, and coal), pollution of land and water due to chemical waste from industries, and its aftermath which includes climate change, deforestation, desertification, ocean acidification and a risk to several species. The term "environmental crisis" refers to the developing state of ecological disequilibrium caused mostly by human activities based on contemporary technology, as expressed by the majority of climatologists. There has been a recent discussion about a specific aspect of the ecological problem, namely global warming. Scientists affiliated with the Intergovernmental Panel on Climate Change (IPCC) discovered inaccurate data on the melting glaciers of the Himalayan, which sparked the following discussion.[14] Notwithstanding, most climate specialists maintain that this finding should not be construed as proof that the crisis is either non-existent or not caused by human activities. Their argument is that the IPCC and other scientific evaluations of climate change, which engage numerous scientists to conduct extensive and thorough investigations, are prone to making errors. However, when these errors are identified, they are corrected. The recent events have not altered the essential conclusions about climate change in any significant way.[15]

Will Stephen and his team, which included Nobel laureate Paul Crutzen, put forth a convincing argument that the current environmental crisis

---

12 Ibid.

13 Peter Doran and Maggie Kendall Zimmerman, "Examining the Scientific Consensus on Climate Change," *EOS* publication of American Geophysical Union, 90, no.3 (January 20, 2009): 22. Naomi Oreskes, "The Scientific Consensus on Climate Change: How Do We Know We're Not Wrong?" in *Climate Change: What it Means for Us, Our Children, and Our Grandchildren,* eds. Joseph Dimento and Pamela Doughman (Cambridge, MA: The MIT Press, 2007), 65-99.

14 Elizabeth Rosenthal, "U.N. Panel's Glacier Warning is Criticized As Exaggerated" New York Times, January 18, 2010.

15 Peter Gleick et al., "Climate Change and the Integrity of Science," Science, vol. 328. no. 5979 (May 7, 2010): 689-690.

can be attributed to unrelenting industrial growth.[16] Their claim was that pre-industrial civilizations had an insignificant effect on the environment and that it is the contemporary economic system and way of life, powered by advanced science and technology, that bears the blame for the environmental decline.[17] Stephen and his colleagues proved that the upsurge in atmospheric CO2 levels commenced solely following the commencement of the industrial revolution. It steadily rose and by 1900, surpassed the highest limits achieved in the preceding 250 thousand years.[18]

In addition, they made the discovery that the acceleration of economic growth and technological advancement correlates with an increase in atmospheric CO2, which currently stands at roughly 33% more than levels present before the industrial era.[19] The exponential growth of the world's population by six times since the Industrial Revolution is frequently considered a leading cause of environmental issues. However, this surge is in fact a direct consequence of the use of fossil fuels to manufacture synthetic fertilizers that employ reactive nitrogen compounds from atmospheric nitrogen. This method has helped to enhance food production and enable the progress of other essential services that paved the way for modern population expansion.[20]

Various novel reports by scientific investigations support the need for a real solution to the situation. In a groundbreaking effort, Johann Rockstrom, who serves as the director of the Stockholm Resilience Centre, collaborated with twenty-six distinguished scientists to define the boundaries that determine the extent to which human society can maintain itself. [21] These experts have recognized nine crucial planetary systems that are at risk due to contemporary lifestyles, including atmospheric CO2 concentration, ocean acidification, stratospheric ozone depletion, nitrogen and phos-

---

16 Will Stephen, Paul J. Crutzen and John R. McNeill, "The Anthropocene: Are Humans Now Overwhelming the Great Forces of Nature," Ambio vol. 36, no. 8 (December 2007): 614-621.

17 Ibid., 615.

18 Ibid., 617

19 Ibid., 618

20 Ibid., 616.

21 Johan Rockstrom et al, "A Safe Operating Space for Humanity," Nature, vol. 461, no. 7263 (September 24, 2009): 472-475.

phorus cycles, global freshwater consumption, alterations in land use, loss of biodiversity, atmospheric aerosol loading, and various forms of chemical pollution. Out of these systems, atmospheric CO2 concentration, nitrogen cycle equilibrium, and biodiversity preservation have already exceeded the acceptable limits. Scientific studies have led to a significant conclusion that our focus cannot be limited to any one planetary system alone, and instead, we must consider the interconnection and interdependence among these systems. If we cross a boundary in one system, it puts other boundaries at serious risk as well.[22]

According to the findings of the Intergovernmental Panel on Climate Change (IPCC), the global temperature is projected to rise by 5.2 degrees Celsius by the end of this century, which is significantly higher than the previous estimate of 2.4 degrees Celsius.[23] Coral reefs may become chemically inhospitable due to the process of ocean acidification caused by the absorption of CO2 by ocean water. This could pose a threat to the habitats of a significant portion of the marine population by the year 2050.[24] Moreover, the quantity of nitrogen compounds that enter the ecosystem through synthetic fertilizers surpasses the amount introduced by all natural processes combined. This has become a significant contributor to greenhouse gas emissions and water pollution.[25] The effects of pollution on both human and non-human life can be profound, even if they are not immediately apparent. In fact, it is predicted that the annual decline in biodiversity will be at least one hundred times greater than the level prior to industrialization, which will have significant consequences for the environment in the coming decades.[26] Studies conducted by David Pimentel, a well-known scientist at Cornell University, indicate that pollution of water, air, and soil is already responsible for forty percent of all global fatalities annually.[27]

---

22 Rockstrom et al, "A Safe Operating Space for Humanity," 474.

23 "The Key Scientific Developments Since The IPCC Fourth Assessment Report of June 2009," prepared by the Intergovernmental Panel on Climate Change (IPCC), in Science Brief 2, June 2009, p. 5, available from www.pewclimate.org.

24 Ibid., 1.

25 Rockstrom et al., "A Safe Operating Space," 474.

26 Ibid.

27 Andrea Thompson, "Pollution May Cause 40 Percent of Global Deaths," LiveScience, September 10, 2007, http://www.livescience.com/environment/070910_pollution_deaths.html.

Many prominent economists, however, have largely rejected these observations about the environment that appear to be relevant to the present economic system. Recently, one individual expressed the belief that "if things go as they have since 1800... The environment will be improving" but this statement emphasizes the need for further discussion on the matter.[28] However, our argument aligns with most climate scientists' views that human actions, facilitated by current knowledge and technology, are the driving force behind the environmental problem.

Over the past few decades, many environmental organizations, as well as national and international economic strategies, have made efforts to address this growing challenge, particularly since the early 1970s. To comprehend Seyyed Hossein Nasr's unique religious and philosophical approach to the issue, we must first examine his background, wide-ranging intellectual work, and his historical position in Western and Islamic environmental activism.

What is New

Despite Nasr's historical significance, the depths of his message, and its present relevance, his response to the environmental crisis has never been presented in a single work comprehensively and systematically. Nasr's investigation into the philosophical roots of the environmental problem leads him to attribute it to rationalism and contemporary technology, as evidenced in his two works: The Encounter of *"Man and Nature"* and *"Religion and the Order of Nature."* As a solution to reinvigorate a religious worldview, he recommends a spiritual reawakening and advocates for the replacement of modern science with a sacred science that honors the metaphysical principles present in all religious traditions. According to Nasr, resolving the environmental crisis in any civilization would necessitate leveraging the conventional intellectual resources of that civilization. Nevertheless, his works do not present a complete and unambiguous plan of action for the resolution he envisions for any given civilization. His approach is evident in several essays and interviews where he discusses

[28] Deirdre N. McCloskey, "Bourgeois Dignity: Why Economics Can't Explain the Modern World" (Chicago: University of Chicago Press, 2010), 12.

how the Muslim world can address the environmental crisis within the present intellectual framework.

In order to grasp Nasr's approach to the Islamic world, it's essential to possess knowledge about various aspects of Islamic intellectual traditions that he alludes to. There is a need for a comprehensive single volume that covers all the aspects of Nasr's perspective on the environmental issue. Furthermore, there is a need for further examination of how Nasr's ideas interact with other environmental philosophies and worldviews, as well as their relevance and applicability to non-Islamic cultures and societies.

This book underscores that Nasr's concurrence is not exclusively centered on physical survival, which is the primary concern of mainstream environmentalism, or moral instructions, as highlighted by religious environmentalism. Rather, it is grounded in his comprehension of the spiritual essence of nature and its relevance to the human mission as elucidated by religion. According to Nasr, it is imperative to have a strong knowledge base because modern science, which he deems accountable for the crisis, contests religions with the knowledge it purports to hold. The expectation is that this study will aid in comprehending this distinguished thinker, who has been advocating spiritual renewal as the ultimate long-term remedy for the environmental crisis since the 1960s. This will be achieved by assembling arguments that are pertinent to the present intellectual climate, a view shared by numerous prominent environmentalists today.

Why this is Study Important

Religion and ecology specialist Max Oelschlaeger emphasized the significance of religion in addressing the environmental crisis. He asserted that some social values such as sustainability and conservation of endangered species and natural habitats cannot be fulfilled through market mechanisms.[29] Rather, religion is the key means for understanding these values as it represents the foremost cultural dialogue outside of the modern narrative of economic expansion and technological solutions. [30] Mary Evelyn Tucker,

---

[29] Max Oelschlaeger, "Caring for Creation: An Ecumenical Approach to the Environmental Crisis" (New Haven: Yale University Press, 1994), 48.

[30] Ibid. p. 47

a distinguished scholar in the field of religion and ecology, also recognizes that religions hold unparalleled moral influence that cannot be found in any other institutions.[31]

Some individuals hold the belief that modern environmentalism, being entirely secular, may not be as effective in persuasion. According to Bryan Appleyard, a notable detractor of modern science and its contemporary applications, the conventional environmental movement only offers a mechanical interpretation of the cosmos. As a result, it falls short of providing any spiritual rationale for the deep-seated change of heart that environmental advocates view as crucial in altering human conduct to address the urgent environmental predicament and stave off its exacerbation:

> "While eco-friendly behavior may bring environmentalists a sense of tranquillity, their underlying motivations are rooted in practical considerations rather than spiritual ones. Their movement is driven by a sense of imminent crisis rather than an aspiration towards a higher understanding. The primary goal is to rectify the damage already inflicted, with no higher ambitions. Even in an imagined "Green paradise," humanity is viewed as subservient to the silent, alien landscape of nature. This perspective echoes the tenets of mechanical determinism, in which nature holds the ultimate power, and humans are forever relegated to a subservient role."[32]

Seyyed Hossein Nasr's perspective on environmentalism is currently gaining importance, particularly due to the lack of a more compelling justification in mainstream environmentalism and the quest to rekindle spiritual or religious principles. Nasr acknowledges that the primary cause of the environmental crisis is the contemporary economic system, which capitalizes on human desires, particularly the greed for material possessions, by manufacturing artificial needs that are not authentic necessities.[33]

---

[31] Mary Evelyn Tucker, "Wolrdly Wonder: Religions Enter Their Ecological Phase" (Chicago: Open Court, 2003), 9.

[32] Bryan Appleyard, "Understanding The Present: An Alternative History of Science" (London: Tauris Parke Paperbacks, 2004), 137.

[33] Seyyed Hossein Nasr, "Religion and the Environmental Crisis," in "The Essential Seyyed Hossein Nasr, ed. William C. Chittick" (Bloomington, IN: World Wisdom, 2007), 31-32.

He queries how we can restrain people from incessantly desiring more without the aid of the Spirit that religion provides, as no other influence in the world holds comparable sway, save brute force.[34] Nasr is a firm believer that religion should be regarded not only as a basis of ethics but also as a fount of knowledge, comprising profound comprehension of God, human existence, and the natural world.[35]

Nasr, who has extensively studied Western culture and spent most of his life in the West, criticizes Western civilization from an Islamic perspective without engaging in "West bashing." He is critical of the secular ideology's tendency to overlook religious ethics and values. While some environmentalists hear him, Western environmental ethicists rarely do. Mainstream ethicists in the field of environmental ethics have shown a preference for secular ethics, while metaphysical or religious ethics are generally not favored.[36] During a Green-Left party conference in the Netherlands, there was a discussion about the significance of ethics in tackling climate change. However, religious ethics were dismissed as "metaphysical" and not given due consideration. Calls for a reconsideration of religion in environmental ethics are occasionally heard but seldom followed up on. The introduction of a 2006 UNESCO publication on Environmental Ethics and International Policy emphasizes the significance of religious environmental ethics, but the volume does not delve into it in depth.[37]

Although a specific ethical stance may be grounded in philosophy, it is believed that religious environmental ethics should receive greater attention. The main reason for this is rooted in pragmatism, as most people around the globe live, work, and think within a religious context. As 'worldviews' and 'value systems,' religions play a pivotal role in guiding human behavior and serve as major sources of inspiration for individuals, communities, and cultures. These ideologies are impossible to disregard. With the global scope and gravity of environmental challenges, it is reasonable to

[34] Seyyed Hossein Nasr, "Religion and the Environmental Crisis," (lecture, Harvard University, Cambridge, MA, February 4, 2008), 32-33.

[35] Ibid., 36.

[36] Marcel Duwell, "Challenges for Individual Rights?," Ethics and Politics of Climate Change conference, Utrecht University, January 23-24, 2009, accessed March 25, 2023, http://www.ethicsandpoliticsofclimatechange.nl/.

[37] "Environmental Ethics and International Policy" (London: Routledge, 2006), 27.

anticipate that an increasing number of Muslims will contemplate these issues from an Islamic perspective in the forthcoming years. There is no inherent conflict between religious ideologies and secular value systems, despite concerns among some in the Western world. At present, various Islamic movements endeavor to establish common ground with other religions, expressing their religious convictions in a universal language that goes beyond the particular cultural customs of Muslims. According to Fuller, an approach centered around values is less intimidating to outsiders compared to an approach that relies heavily on strict doctrine or ritual. Nasr's eco-philosophy embodies this value-centered approach to addressing the shared challenges faced by communities around the world, including the environmental crisis. In America, many Muslim communities and mosques are actively engaging with non-Muslim members to seek common solutions to shared problems.[38]

Thus study on Seyyed Hossein Nasr's environmental philosophy is necessary for several reasons:

1. Nasr is one of the foremost scholars of Islamic environmental thought, and his work has made a significant contribution to the development of this field. Researching his ideas can help to deepen our understanding of Islamic environmental ethics and its potential for promoting sustainability and environmental justice.

2. Nasr's environmental philosophy offers a critique of modernity and its impact on the environment, which is relevant to contemporary debates in environmental ethics and environmental policy. His ideas can help us to rethink our relationship with the natural world and develop more sustainable and just ways of living.

3. Nasr's work engages with multiple intellectual traditions, including Islamic, Sufi, and Perennialist thought. Studying his ideas can help to promote interfaith and cross-cultural dialogue on environmental issues and highlight the role of religious and

---

[38] Fuller, G. "*The Future of Political Islam,*" New York: Palgrave MacMillan, 2003, p.205.

spiritual perspectives in shaping our relationship with the natural world.

4. Nasr's environmental philosophy has implications for a wide range of fields, including environmental ethics, philosophy, theology, and cultural studies. Researching his ideas can help to bridge disciplinary divides and foster interdisciplinary collaboration in addressing environmental challenges.

Overall, research on Seyyed Hossein Nasr's environmental philosophy can help to enrich our understanding of the relationship between religion, culture, and the environment and inspire new approaches to environmental ethics and sustainability.

Failures of Mainstream Environmentalism Today

In the United States during the early 1960s, the public became increasingly aware of the environmental issues that were affecting their local communities. This was due to the fact that the country was heavily industrialized, making it more likely for people to notice the pollution that resulted from industrial activities. As time passed, it became clear that modern industrialization was the main reason for the ecological problem that needed to be addressed. In 1972, the publication of the Club of Rome's Limits to Growth brought the issue of environmental protection to the forefront of public consciousness.[39] Consequently, a number of environmental protection laws were passed by 1976. By doing so, the intersection of business interests, social movements, and political movements portrayed environmentalists as individuals who overreacted to situations, creating a negative image of them. During the 1980s and 1990s, advocates of economic growth devised a fresh approach to weaken the environmental movement, particularly as the worldwide scope of the crisis became increasingly apparent. During the early 1960s, environmental issues at the local level began to be acknowledged by the United States, and by the early 1970s, it was commonly accepted that contemporary industrialization was the primary contributor to environmental deterioration. Although laws were enacted to support

39 Donella H. Meadows, Dennis L. Meadows, Jorgen Randers, and William W. Behrens III, "The Limits to Growth" (New York: Universe Books, 1972),

environmental protection, they inadvertently led to the marginalization of the environmentalist movement, allowing business interests and other groups to dismiss environmental concerns as alarmist.

In his book "Planet Dialectics," Wolfgang Sachs presents a critical analysis of environmental policies aimed at addressing the issue of environmental degradation.[40] According to Sachs, these policies have been primarily focused on managing the remaining natural resources of the world, without challenging the underlying economic goal of perpetual growth.

Sachs identifies two key components of these policies. The first component is the belief that technological advancements will enhance efficiency and reduce the number of resources required for a given result of performance. This idea implies that as efficiency increases, the negative impacts of economic development will be offset. However, Sachs argues that this notion fails to address the underlying problem of resource depletion and does not challenge the assumption of perpetual growth.[41] The second component of these policies, as Sachs sees it, is the argument that poverty in developing countries drives people to exploit natural resources. This argument has led policymakers to push for economic growth to alleviate poverty. Sachs suggests that this approach does not address the underlying problem of resource depletion and instead perpetuates the cycle of exploitation. Sachs argues that environmental policies need to challenge the dominant economic paradigm of growth and focus on sustainable management of natural resources. Such policies need to address the root causes of environmental degradation and poverty, rather than just their symptoms.[42]

In recent times, economic policymakers have ramped up developmental efforts in the name of safeguarding the environment from impoverished communities with the aim of lifting them out of poverty. However, in doing so, they overlook the fact that such actions often exacerbate ecolog-

---

[40] Wolfgang Sachs, "Planet Dialectics: Explorations in Environment and Development" (Halifax, Nova Scotia: Fernwood Publishing, 2001),

[41] Ibid., 47-48, 59.

[42] Ibid., 60. Lloyd Timberlake, "The Emergence of the Environmental Awareness in the West," in "The Touch of Midas," ed. Z. Sardar (Manchester Univ. Press, 1984), 127.

ical scarcity. In reality, the individuals who have always relied on the environment have no alternative but to pursue the last remaining vestiges of its resources. [43] Additionally, the term "development" and the expression "poverty reduction" have acquired a sacred status in economic terminology, with their definitions deemed beyond criticism. This has led to project proposals that are filled with such terminology to make them more appealing for funding. Unfortunately, real questions about what is actually done are often left unanswered. As observed by Andrea Cornwall, the economic rhetoric has been carefully crafted to imbue these phrases with a sense of nobility, while obscuring the negative consequences of these actions.[44] It is crucial to recognize that impoverished communities who depend on the environment have intrinsic knowledge of how to manage natural resources sustainably. Therefore, it is imperative to involve them in the decision-making process when designing and implementing development projects. This will ensure that the initiatives are not only effective in reducing poverty but also ecologically sustainable, benefiting both people and the environment.[45]

Sachs argues that while pursuing technical efficiency to reduce environmental harm may offer some advantages, it can also reinforce the economic objective of ongoing growth in production and consumption within a limited global system, resulting in potential drawbacks over time. Ironically, a growing economy's faster development in demand ignores the cost-saving effects of greater efficiency.[46] In fact, several studies conducted over the last decade have found that despite technological advancements, resource usage and pollution have not usually decreased.[47] In general, increased productivity and industrial waste have followed increases in efficiency. Because of this, efficiency efforts will not be successful in reducing resource use as long as economic development continues. Sachs notes that

---

43 Wolfgang Sachs, "Planet Dialectics: Explorations in Environment and Development" (Halifax, Nova Scotia: Fernwood Publishing, 2001), 35.

44 Andrea Cornwall, "Buzzwords and fuzzwords: deconstructing development discourse," Development in Practice, Volume 17, Numbers 4-5, August 2007, p. 472.

45 Ibid.

46 Sachs, "Planet Dialectics", 41-42.

47 James Gustave Speth, "The Bridge at the Edge of the World" (New Haven: Yale University Press, 2008), 56-57.

regions that have already attained a certain level of industrialization are the only ones where seeking efficiency may be most effective.[48] If not for that, the utilization of modern technology can be justified by increased productivity alone. Simply prioritizing efficiency in material production overlooks the non-material goals and principles connected to "less efficient" traditional production techniques.[49]

In the 30-year update of the 1972 classic Limits to Growth, the authors observed that the adoption of efficient technologies takes time, money, and effort. Technologies are often designed to solve specific problems, and their development is a slow and costly process. [50] Even after proving their viability in laboratories, further delays are caused by the need to develop necessary resources such as capital, labor, sales, service staff, marketing, and finance mechanisms required for widespread use. [51] Although favorable market forces, political will, and societal support can help expedite the adoption of efficient technologies, it still takes several years. In a race against time, such as the current environmental crisis, this delay can significantly reduce their effectiveness. Therefore, it is crucial to address environmental problems promptly and invest in the research and development of efficient technologies that can be brought to the market more quickly.

Meadows et al. predict that if the world does not take significant steps to reduce its environmental impact, various aspects of life may improve until 2025, but then rapidly decline.[52] The decline would be due to a lack of access to non-renewable resources. Even under optimistic assumptions, such as a 100% increase in accessibility to non-renewable resources and a 4% annual improvement in pollution-reducing technology, the decline would only be delayed by about 50 years.[53] [54] The primary discovery of

[48] Sachs, "Planet Dialectics", 41.

[49] Ibid., 51, 53-54.

[50] Donella Meadows, Jorgen Randers, and Dennis Meadows, "Limits to Growth: The 30-Year Update" (White River Junction, Vermont: Chelsea Green, 2004), 471.

[51] Ibid., 212.

[52] Meadows et al., Ibid., 168-169.

[53] Speth, James Gustave. "Bridge at the Edge of the World: Capitalism, the Environment, and Crossing from Crisis to Sustainability" (New Haven, CT: Yale University Press, 2008), 114.

[54] Meadows et al., "Limits to Growth: The 30-Year Update" (White River Junction, VT: Chelsea Green Publishing, 2004), 210-11.

the authors is that if society continues to delay confronting environmental constraints by means of economic and technological adjustments, it will face a greater likelihood of confronting several limits simultaneously. This finding underscores the importance of taking proactive measures to limit economic development to avoid exacerbating environmental challenges."[55] Meadows et al.'s research highlights the need for sustainable development practices that balance economic growth with environmental protection. It emphasizes the importance of recognizing that natural resources are finite and that efforts to improve efficiency and productivity must be accompanied by efforts to reduce consumption and waste. By acknowledging the limitations of our current economic systems and taking steps to address them, society can work towards a more sustainable future.

Sachs argues that a mechanistic perspective of the environment has led to a reliance on technological solutions to environmental issues. Scientific empiricists redefined ecology in the early 20th century as a self-regulating environment, diminishing the concept of interdependence among natural elements. [56] The term "homoeostasis" is now used to describe this self-regulation, relying on mechanical engineering principles. As a result, environmentalists have difficulty viewing ecology beyond a mechanical framework. [57] Sachs suggests that a broader understanding of ecology is needed, one that recognizes the interconnectedness of all natural systems and is not solely dependent on technology to solve environmental problems.[58]

The ecosystem has been viewed as a mechanical system that can be understood and managed scientifically. This perspective portrays ecology as a functional machine without a higher goal. Jane Bennett, a renowned political theorist, notes that modern environmentalists use techniques to rationalize nature and transform it into well-managed residential, recreational, commercial, and industrial sites.[59] Sachs contends that modern

---

55 Meadows, Donella H., Dennis L. Meadows, Jorgen Randers, and William W. Behrens III. "The Limits to Growth" (New York: Universe Books, 1972), 223.

56 Sachs, Wolfgang. "Planet Dialectics: Explorations in Environment and Development." London: Zed Books, 1999, 62-63.

57 Ibid., 63.

58 Ibid., 62.

59 Bennett, Jane. "Unthinking Faith and Enlightenment: Nature and the State in a Post-Hegelian Era." New York: New York University Press, 1987, 47.

environmentalists seek to establish the maximum amount of harm an ecosystem can endure before collapsing and how to sustain that equilibrium by manipulating technology, all while ignoring the underlying causes of environmental degradation.[60] Environmental management often focuses on setting threshold values for pollutants that must not be exceeded, according to Bennett. However, the preference for technical solutions among environmentalists is partly due to opposition from industries and corporations that resist pursuing wider or more fundamental environmental goals.[61] The focus on technology as a solution to environmental problems disregards the interconnectedness of natural systems and the need to address the underlying causes of ecological damage. To create a sustainable future, there is a need to develop a comprehensive understanding of ecology that recognizes the value of natural systems beyond their functional purposes and the importance of addressing developmental aspirations that harm the environment. It is crucial to adopt a holistic approach to environmental management that balances economic development with ecological sustainability.

James Gustave Speth, a well-respected environmental advisor who held top positions in the Carter administration, founded the World Resources Institute and was the head of the Yale School of Forestry and Environmental Studies, has determined that in various ways, conventional environmentalism has been unsuccessful in preserving the environment. In his book, The Bridge at the Edge of the World, Speth admits that he believes the current methods and styles of environmentalism are too limited, despite being well-intentioned.[62] As James Halteman notes, Speth's reputation as a leader in environmental reform makes his book a "confession." [63] Speth argues that mainstream environmentalists are working within a restricted framework to protect what they can, but that this approach is not sufficient. Rather, a more comprehensive approach that addresses the underlying systemic causes of environmental problems is necessary.

---

60 Sachs, Wolfgang. "Planet Dialectics: Explorations in Environment and Development." London: Zed Books, 1999, 36.

61 Bennett, Jane. "Unthinking Faith and Enlightenment: Nature and the State in a Post-Hegelian Era." New York: New York University Press, 1987, 47.

62 Speth, James Gustave. "The Bridge at the Edge of the World: Capitalism, the Environment, and Crossing from Crisis to Sustainability." New Haven: Yale University Press, 2008, 78.

63 Halteman, James. "A Sustainable World." Christian Century, December 16, 2008, 38.

Speth has emphasized that environmentalists who work within the existing framework face significant limitations, particularly due to the foundation of modern capitalist culture which is driven by science and technology. According to him, this society fosters a continuous increase in environmental harm as influential companies, with insufficient transparency and limited supervision, employ technology to pursue financial gains and expansion. This principle creates a significant impediment to environmental protection efforts. [64]

Environmentalists face significant challenges when dealing with corporations and industries due to the latter's economic interests. Moreover, the complexity of technologies needed to identify and restore polluted environments is daunting. Consequently, a formidable regulatory and management system has emerged in response to the resistance from commercial and industrial sectors, which is challenging for both laypeople and experts to comprehend. The struggle against corporate opposition often forces environmentalists to concentrate on reducing the visible symptoms of environmental damage by striving for greater technical proficiency instead of addressing the underlying causes of environmental crises.[65] [66] Despite their efforts, these challenges remain a significant impediment to their work, requiring further exploration and possible solutions. [67]

The Call for Spiritual Values

The environmental activists we addressed above, as well as many others, believe that the resurgence of what we may define as spiritual values is a prerequisite for any viable plan that would allow mankind to escape the environmental crises from a deteriorating state.

While Speth considers scientific progress as one of the most important instruments for dealing with the situation,[68] he admits that embracing modern technological solutions to fulfil our material demands has led us to

64 Ibid., 82.

65 Ibid., 83.

66 Ibid., 84.

67 Ibid., 85.

68 Ibid., 113

the ecological crisis we confront today: We developed a potent technology and established an institution to deploy it swiftly and effectively, even at the expense of others if necessary. As a result, we managed to dominate nature and generate unprecedented prosperity. However, we became so enamored and dependent on these systems and achievements that we ignored warning signals along the way. We must refocus on being rather than having, giving rather than receiving, prioritizing needs over wants, and striving for improvement rather than material wealth. [69] Speth believes that altering our perspective is crucial to addressing the challenges of our time. Given that religious doctrines have long emphasized the values of simplicity, generosity, and mindfulness, he suggests that faith communities could play a pivotal role in this endeavor. He points to the example of Dr Martin Luther King Jr., whose Christian faith propelled him to challenge racial injustice and advocate for God's justice. Speth maintains that our generation must follow in King's footsteps and harness the transformative power of faith to effect meaningful change.[70]

The authors of "Limits to Growth" have drawn conclusions similar to Speth's after over thirty years of seeking solutions to the crisis. They also recognize the importance of modern technology but argue that a significant shift in consciousness is required to direct resources towards developing sustainable technologies. While impressive technological advancements are possible, they can only occur through determined societal decisions and a willingness to act and invest. [71] Such decisions, they contend, can only be made following a "structural change" in dominant consciousness, which currently views individuals mainly as producers and, consumers and associates social position with material accumulation. This mindset prioritizes acquiring more, rather than giving more or having enough. Failure to change this mindset will result in technologies and markets that further harm the environment.[72] The authors argue that the necessary change in consciousness is advocated in nearly every religious text. It involves changing people's goals, purpose, and understanding of their place in the world. This change is not political or physical, but internal – in

[69] Ibid., 236.

[70] Ibid., 232.

[71] Meadows et al., "Limits to Growth", 204.

[72] Ibid., 223-24.

people's heads and hearts. They suggest that a shift towards being, rather than having, is necessary to achieve this change. This means prioritizing sustainability, community, and social justice over material wealth and accumulation. Overall, they emphasize that societal decisions and actions must be driven by a fundamental shift in consciousness if we are to develop sustainable technologies and prevent further harm to the environment.[73]

According to Sachs, modern environmentalism has failed to address the environmental crisis as a civilizational problem, instead treating it as a technical issue. This is because the level of productive performance achieved by modern society is no longer viable, not just in the North but globally. Sachs suggests that the missing components in all environmental strategies are "enlightened restraint" and "intelligent self-limitation" of production and consumption. In order to tackle this problem, he suggests turning to the wise teachers of both the East and the West, who have advocated for the principle of simplicity in life for a long time.[74] [75] [76] His recommendation is to refer to the teachings of wisdom from both the East and the West, which have almost unanimously promoted the idea of leading a simple life. This principle requires a limited but skillful use of material objects. While it is true that many of the teachers of wisdom who have endorsed this principle have been devoted followers of a particular religion, Sachs maintains that this approach is applicable to all.[77] Notable figures who have championed simplicity in modern times include Mahatma Gandhi, Mother Teresa, the Dalai Lama, and Sufi Masters like Shaykh Ahmad al-'Alawi and Shaykh Muhammad Rahim Bawa Muhaiyaddeen. To address the environmental crisis, we must address the deeper cultural and social factors that underlie it and make a fundamental shift in our values and priorities. By embracing simplicity, we can create a more sustainable and just future for all.

Email: mhimran@tongji.edu.cn; imran.du24@gmail.com;
muzzamelhussainimran@gmail.com

ORCID ID: https://orcid.org/0000-0001-8141-6654

---

[73] Ibid., 240.

[74] Sachs, "*Planet Dialectics*," 68.

[75] Ibid., 41, 48

[76] Ibid., 49,67

[77] Ibid., 212.

Chapter 1

Nasr's Writings and Basic Thoughts

This section serves as an introduction to the esteemed Islamic scholar and environmental philosopher, Seyyed Hossein Nasr. Nasr's extensive contributions to Islamic philosophy, Islamic science, Sufism, Islamic art, and the school of perennial philosophy have earned him widespread recognition as one of the most influential Islamic philosophers of contemporary times. His remarkable career as an academic and administrator has spanned over five decades, with notable accomplishments in Iran, Lebanon, and the United States.

Our focus in this chapter will be on Nasr's perspective on the ecological crisis, which is centered around the restoration of the religious outlook and the rejection of scientism. Nasr posits that the environmental problem is a result of a secular and materialistic mindset that ignores the spiritual dimensions of nature. His approach seeks to bridge the gap between science and religion by advocating for a holistic and integrated approach to the environment.

Additionally, we will delve into some of Nasr's notable works, which encompass a diverse range of subjects relating to Islam, modernity, and spirituality. His writings are characterized by a deep reverence for traditional wisdom, a critique of the excesses of modernity, and a commitment to promoting the spiritual dimension of human life. Overall, this chapter aims to present a comprehensive overview of Nasr's contributions to Islamic thought and environmental philosophy.

Nasr's life and works

Seyyed Hossein Nasr, a highly esteemed academician, was born in Tehran in 1933 into a family of intellectuals. His father, Seyyed Valiullah Khan Nasr, was a distinguished philosopher with a deep understanding of both Arabic and Persian languages, and his mother was also an accomplished scholar. It was in this academic environment that Nasr's interest in the nature of physical reality was sparked, leading him to pursue undergraduate studies in physics at the Massachusetts Institute of Technology (MIT) in 1950.[78] After graduating from MIT in 1954, Nasr went on to pursue a PhD in the history of science at Harvard University, where his studies became more closely linked with philosophy. During the course of his studies, Nasr developed a profound appreciation for the historical and cultural context in which science developed, which would later influence his work as an academic and administrator. Over the past five decades, Nasr has made significant contributions to the fields of philosophy, religion, and science. He has held academic positions at numerous institutions in Iran, the USA, and Lebanon and has authored over fifty books and hundreds of articles on topics ranging from metaphysics to Islamic art and architecture. Since 1984, Nasr has served as a distinguished professor of Islamic Studies at George Washington University's Department of Religion in Washington, D.C. His exceptional scholarship and administrative abilities have earned him widespread recognition, and he continues to inspire new generations of students and scholars with his intellectual curiosity and academic rigor.

Seyyed Hossein Nasr is widely considered to be one of the most influential Islamic intellectuals of our time. He has produced a vast collection of written material that comprises almost fifty books and hundreds of articles on a wide range of subjects, encompassing Islamic philosophy, science, art, the ecological problem, Sufism and the school of perennial philosophy propagated by René Guénon (1886-1951) and Frithjof Schuon (1907-98). Throughout his long and illustrious career, Nasr has earned numerous accolades and achievements. Perhaps the most prominent among these is his selection as the distinguished Gifford Lecturer at the University of Edinburgh in 1981, a prestigious honor that is bestowed upon individuals who have made significant contributions to the field of natural theology. In 2001, Nasr was also included in the Library of Living Philosophers, a distinguished list that recognizes the most influential philosophers of our time.[79] His work has earned the admiration and respect of scholars across

[78] Seyyed Hossein Nasr. "An Intellectual Autobiography." In "The Philosophy of Seyyed Hossein Nasr", edited by Lewis Edwin Hahn, Randall E. Auxier, and Lucian W. Stone Jr., The Library of Living Philosophers, Volume XXVII, 16. Chicago and La Salle: Open Court Publishing Company, 2001.

[79] "The Philosophy of Seyyed Hossein Nasr", edited by Lewis Edwin Hahn, Randall E. Auxier, and Lucian W. Stone Jr., The Library of Living Philosophers, Volume XXVII (Chicago and La Salle: Open Court Publishing Company, 2001), introductory pages.

disciplines. Huston Smith, a prominent intellectual of world religions, has described him as "one of the major intellects of our day," while Keith Critchlow, a leading scholar of religious art and architecture, has dubbed him "the most important living philosopher on the planet today."[80] [81]Nasr's impact on Islamic thought has been immense, as he has sought to bridge the gap between traditional Islamic scholarship and the modern world. He has argued for the importance of understanding the historical and cultural context in which Islamic thought developed, and has advocated for a renewed focus on traditional Islamic sources in contemporary Islamic discourse. Through his writings and teachings, Nasr has inspired countless students and scholars to explore the rich and complex traditions of Islamic thought and to engage in critical reflection on the challenges facing the contemporary world. His influence continues to be felt in academic and intellectual circles around the globe, making him one of the most important and respected figures in the contemporary Islamic world.

Nasr not only holds the distinction of being the first Islamic and Muslim scholar to address the ecological crisis, but he has also authored numerous works on the philosophical and religious aspects of this matter.[82] [83] Nasr is notable for being the pioneer in advocating for a holistic and all-encompassing religious solution that surpasses mere ethical considerations, setting him apart from all previous individuals. As the foremost spokesperson of the perennial philosophy school, he has presented the most comprehensive and exhaustive approach to the environmental crisis among all other proponents of this school.[84] Nasr's environmentalism emphasizes

---

80 Huston Smith. Quoted on the dust jacket of Seyyed Hossein Nasr, "Religion and the Order of Nature" (New York and Oxford: Oxford University Press, 1996).

81 Keith Critchlow. "Keyonte Speech." In "The Beacon of Knowledge: Essays in Honor of Seyyed Hossein Nasr," edited by Mohammad H. Faghfoory, xlvii. Louisville, Kentucky: Fons Vitae, 2003.

82 Richard Foltz. "Introduction to Islam and Ecology: A Bestowed Trust." Edited by Richard C. Foltz, Frederick M. Denny, and Azizan Baharuddin, xxxviii. Cambridge, MA: Harvard University Press, 2003.

83 Giovanni Monastra. "Seyyed Hossein Nasr: Religion, Nature, and Science." In "The Philosophy of Seyyed Hossein Nasr," edited by Lewis Edwin Hahn, Randall E. Auxier, and Lucian W. Stone Jr., The Library of Living Philosophers, Volume XXVII, 496-7. Chicago and La Salle: Open Court Publishing Company, 2001.

84 Christopher James, the 5th Lord Northbourne, and Joseph Fitzgerald, eds., "Of the Land and the Spirit: The Essential Lord Northbourne on Ecology" (Bloomington, IN: World Wisdom, 2008).

the need to revive religiously-centered lifestyles globally as a substitute for the modern lifestyle that is driven by technology and economic interests, which he attributes as the cause of the ecological crisis. This comprehensive feature of Nasr's philosophy, as we will delve into, mirrors the essence of both the perennial philosophy school and the core aspect of Islam.[85]

To gain insight into Nasr's approach to the ecological problem, it is essential to grasp the meaning of three critical terms: 1) Religious viewpoint, 2) Scientific viewpoint, and 3) Scientism. These concepts can be succinctly explained as follows: 1) Religious viewpoint pertains to the fundamental nature of the universe as perceived by a religion, which holds that the cosmos is replete with indications of the divine.[86] 2) According to the scientific viewpoint, the universe is made up of ontologically independent material phenomena that do not have any connection to a supernatural source of their existence. This is how current science sees the nature of reality in the universe.[87] 3) Nasr contends that scientism, which upholds modern science as the preeminent or perhaps only dependable means of attaining genuine knowledge, has been the prevailing fundamental doctrine in the West and other Westernized area of the world since the 17th-century Scientific Revolution, wherein the scientific perspective gradually supplanted the religious viewpoint.[88] [89] Nasr argues that the present condition is the origin of the environmental predicament and demands transformation. The restoration of the religious outlook is vital for the durable resolution of the environmental dilemma. Although the reliance on scientific research seems to conflict with holding advanced technology accountable for the ecological problem, this is because Nasr's stance does not refute the ability of modern science to comprehend the physical aspect to a considerable extent.[90] The

---

85 Seyyed Hossein Nasr, "The Need for a Sacred Science" (New York: State University of New York, 1993), 71-94.

86 Mircea Eliade, "The Sacred and the Profane: The Nature of Religion, trans. Willard Trask" (New York: Harcourt, 1987), 116.

87 John J. Carvalho IV, "Overview of the Structure of a Scientific Worldview," Zygon 41, no. 1 (March 2006): 113-124.

88 Huston Smith, "Scientism: The Bedrock of the Modern Worldview," in "Science and the Myth of Progress, ed. Mehrdad M. Zarandi" (Bloomington, IN: World Wisdom, 2003), 233.

89 Huston Smith, "Beyond the Postmodern Mind : The Place of Meaning in a Global Civilization" (Wheaton, IL: Theosophical Publishing House, 2003), 244.

90 Seyyed Hossein Nasr, "Islam and the Problem of Modern Science," in "An Ear-

complication with modern science, as per Nasr, is its refusal or neglect of the existence of any actuality beyond nature's material dimension.[91]

Nasr's contribution to academic literature is remarkable, as he boasts a prolific publication record that spans several decades. With an impressive collection of over 50 books and hundreds of articles, he has firmly established himself as a leading authority in various fields of study, including Islamic cosmology, theology, science, metaphysics, Sufism, philosophy, religious pluralism, Islamic art and architecture, Persian mysticism, and modernity. Apart from his original works, Nasr has also edited several influential anthologies and encyclopedias, which have been instrumental in shaping our understanding of these complex subjects. Some of his notable works in this regard include "Islamic Spirituality" (1991)[92], which offers an in-depth analysis of the spiritual traditions in Islam; "History of Islamic Philosophy" (1996) [93], " a comprehensive survey of Islamic philosophical thought from its origins to the modern era; "An Anthology of Philosophy in Persia" (1999, 2000),[94] which explores the rich intellectual legacy of Persian philosophy; and "The Heritage of Sufism" (1999) [95], a seminal work that delves into the mystical dimensions of Islam.

In addition to his scholarly pursuits, the author has produced a significant number of works intended for a general audience, which aim to introduce or examine the relationship between Islam and modernity. His publications include "Traditional Islam in the Modern World," which was first published in 1985, "Islam and the Plight of Modern Man," published in 1975, and "A Young Muslim Guide to the Modern World," which was published in 1998.

---

ly Crescent: The Future Knowledge and Environment in Islam, ed. Ziauddin Sardar" (London: Mansell, 1998), 133.

91 Seyyed Hossein Nasr, "The Need for a Sacred Science" (New York: State University of New York, 1993), 4.

92 Nasr, S.H., ed. "The Encyclopaedia of Islamic Spirituality" (Lahore: Suhail Academy, 2000).

93 Nasr, S.H. and Leaman, O., eds. "History of Islamic Philosophy" (London and New York: Routledge, 1996), 2007 of the Routledge History of World philosophies.

94 Nasr, S.H. and Amanirazavi, M., eds. "An Anthology of Philosophy of Persia" (Lahore: Suhail Academy, 2005).

95 Lewisohn, L., ed. "The Heritage of Sufism" (Oxford: Oneworld, 1999).

These works have gained widespread recognition and have been translated into an impressive 22 languages, including Japanese, Indonesian, Turkish, Bosnian-Serbo-Croatian, Persian Arabic, Polish, Urdu, Dutch, French, Tamil, and others, making them accessible to a broad range of readers across the world. Nasr's most recent books, "The Heart of Islam: Enduring Values for Humanity," published in 2004, and "The Garden of Truth: The Vision and Promise of Sufism, Islam's Mystical Tradition," published in 2007, aim to provide a more favorable view of Sufism and Islam and to non-Muslim readers. These books offer a deep insight into the mystical and philosophical dimensions of Islam, focusing on its spiritual and moral values and their relevance in the contemporary world.

Nasr's work is centered around Islam, but also has a comparative approach known as "perennials." He explores and compares major philosophical and metaphysical systems in his work. One of his key concepts is "tradition." Nasr is a member of the "Traditionalist school" and his philosophy can be found throughout his work, particularly in his most significant work, "Knowledge and the Sacred" (1989). The book was based on a series of lectures at the University of Edinburgh, known as the Gifford Lectures, which was delivered by him in 1981. This was considered a turning point in Nasr's career. After leaving Iran, he wrote the lectures and they crystallized in his mind like a "running river."[96] Soon after, he continued his academic career at George Washington University.

> I have multiple objectives. Initially, I strive to delve into the diverse aspects of natural sciences, traditional metaphysics, and cosmologies that contrast with contemporary science. Secondly, I strive to discover alternative methods for studying the history of science beyond the dominant positivist approach, drawing inspiration from the works of Pierre Duhem. Thirdly, I endeavor to establish a genuine methodology for studying Islamic science and philosophy from an internal perspective. Fourthly, I seek to revive the entirety of the Islamic philosophical tradition, including, philosophy, Sufism, sciences, and arts, within the context of contemporary society.
>
> Additionally, I aspire to approach Western thought from the

[96] Seyyed Hossein Nasr, "The Philosophy of Seyyed Hossein Nasr" (2001), 77-78.

> perspective of the Islamic philosophical tradition. Finally, I am committed to engaging in profound intellectual and philosophical discourse that addresses the fundamental tensions between the East and West, as well as between tradition and modernity.[97]

Nasr's expertise is not limited to Islamic philosophy; rather, he is recognized as a philosopher in his own right. According to Majid Fahkry, Nasr is regarded as the most distinguished Iranian philosopher of modern times. Nasr has authored numerous works on topics such as mysticism, Islamic metaphysics and cosmology, and is held in high esteem within academic communities.[98] Nasr was featured in The Library of Living Philosophers series in 2000, a compilation that also included distinguished philosophers such as Buber, Sartre, Dewey, and Popper. Given his active participation in interfaith and intercultural dialogue, Nasr's work holds immense significance in our present times.[99]

Nasr's passion for the environmental crisis is a significant aspect of his philosophical interests. In 1958, he presented his doctoral book on traditional thoughts of nature and cosmologies, which he later published as "An Introduction to Islamic Cosmological Doctrines" in 1993. "Religion and the Order of Nature," published in 1996 [100] is his most recent work on this topic and is considered the culmination of his series on the interrelationship between religion, science, and nature, which he has consistently written about since the late 1960s.[101] [102]

The Rockefeller Series Lectures were a seminal work that delved into the complex relationship between humans and the natural world. Initially published in 1966 as "The Encounter of Man and Nature," the collection of essays was later reprinted in 1987 as "Man and Nature: the Spiritual Crisis of Modern Man." These works offer a comprehensive and comparative

---

97 Seyyed Hossein Nasr, "The Philosophy of Seyyed Hossein Nasr" (Autobiography), (Tehran: Soroush Press, 2001), 77-78.

98 Ibid.

99 Ibid.

100 Nasr, S.H., "Religion and the Order of Nature" (New York, Oxford: Oxford University Press, 1996), page number(s) if applicable.

101 "Reply to A.K. Saran," in "The Philosophy of Seyyed Hossein Nasr" (2001), 441.

102 *"The Philosophy of Seyyed Hossein Nasr"* (till 1999), the internet, references in other publications, etc..

analysis of different religious traditions and their perspectives on environmentalism and ecological crises.[103] [104] Of particular note is the attention paid to the Islamic perspective in these works. The author explores the Islamic view of nature and the environment in "Sufi Essays," published in 1972. This work focuses on Sufism, the mystical branch of Islam, and its approach to the ecological crisis. The author examines how Sufism's spiritual and ethical teachings can provide guidance on the preservation of the natural world. The author also devotes considerable attention to the Islamic perspective on the relationship between knowledge and spirituality in "Knowledge and the Sacred," published in 1989,[105] and its sequel "The Need for a Sacred Science," published in 1993. These works delve into the Islamic tradition's emphasis on the unity of knowledge and the interconnectedness of all things. The author's exploration of Islam in these works is particularly noteworthy as it provides a unique perspective on the intersection of religion and ecology [106], Later, the author reworked a chapter from "The Need for a Sacred Science" into a standalone essay titled "Islam and Ecology: a Bestowed Trust," published in 2003 as part of the Harvard Divinity School series on Religion and Ecology.[107] In this work, the author examines the concept of "*amanah*" in Islamic theology, which refers to the responsibility and trust bestowed upon humans by God to care for the environment. Nasr is often invited to speak about his subject and has been a guest speaker at events such as "Nature and the Sacred: A Fierce Green Fire" a conference at Oregon University where he addressed deep ecologists and eco-spirituals in 2004.[108] He also spoke at the San Rossore

---

[103] Nasr, S.H. "The Encounter of Man and Nature: The Spiritual Crisis of Modern Man." London: Allan and Unwin, 1968. Reprinted as Man and Nature (1987).

[104] Nasr, S.H. "Man and Nature: The Spiritual Crisis of Modern Man." London: Unwin Paperbacks, 1987. New edition: London: Unwin and Hyman and Harper-Collins, 1990. Chicago: ABC International Group, 1997.

[105] Nasr, S.H. "Cosmos as Theophany." In "Knowledge and the Sacred," by S.H. Nasr, 87-105. Albany: State University of New York Press, 1989.

[106] Nasr, S.H. "Sacred Science and the Environmental Crisis: An Islamic Perspective." In "The Need for a Sacred Science", by S.H. Nasr, 163-176. Albany: State University of New York Press, 1993.

[107] Nasr, S.H. "Islam, the Contemporary Islamic World, and the Environmental Crisis." In "Islam and Ecology: A Bestowed Trust", edited by R.C. Foltz, F. Denny, and A. Baharuddin, 73-87. Cambridge, Massachusetts: Harvard University Press, 2003.

[108] Conference on Nature and the Sacred, "A Fierce Green Fire," (October 2004), http://springcreek.oregonstate.edu/nature_sacred/index.html.

conference that year to the environmental sector. In recent years, Nasr has had a conversation with Muzaffar Iqbal, an advocate of Islamized science, aimed at Muslim audiences.[109] Nasr continues to address various audiences.[110]

Nasr was one of the first in the Western world to bring attention to the ecological issue. Although Rachel Carson's "Silent Spring" (1962) is considered the first work to raise the ecological problem in the public consciousness, Nasr's own book on the topic followed. The impact of "Silent Spring" reinforced Nasr's belief in an impending disaster and inspired him to investigate its causes. Nasr, in his twenties at the time, concluded that the environmental crisis resulted from a spiritual crisis within modern man, rather than exclusively from engineering problems.[111] He arrived at this realization after witnessing the destruction of vast, pristine forests during the 1950s.

> My concern with environmental issues dates back to the early 1950s when I was a student at M.I.T. and Harvard University. Even then, I was acutely aware of the beauty of nature and would often take solitary walks around Walden Pond, much like Thoreau. However, it wasn't until the construction of Route 128 around Boston that I realized something was fundamentally wrong with our relationship with nature. The beltway's construction ecologically separated the area inside from the relatively unspoiled countryside beyond, highlighting the crisis in our relationship with the natural world. I foresaw a spiritual crisis looming on the horizon, with the root causes of the environmental catastrophe deeply ingrained in our modern industrial practices. From my perspective, it appeared as if nature was harboring cancer that would eventually disrupt its delicate balance and harmony, resulting in its demise as we currently recognize it.[112]

---

[109] Seyyed Hossein Nasr, "The Islamic perspective on the environmental crisis: Seyyed Hossein Nasr" in "conversation with Muzaffar Iqbal," accessed October 2008, www.thefreelibrary.com.

[110] Hamza Yusuf, "Faith and the Environment," public lecture, Revival Series, May 17, 2008, http://www.revivingtheislamicspirit.com/.

[111] Seyyed Hossein Nasr, "Reply to Giovanni Monastra," in The Philosophy of Seyyed Hossein Nasr (2001): 516.

[112] R.C. Foltz, ed., "Islam and Ecology: A Bestowed Trust" (2003), 85-86.

> Since my early years of life, I developed a deep affection for the diverse manifestations of nature. I was captivated by the allure of towering mountain peaks, which held a mystical sway over my imagination. I also found myself enchanted by the sprawling expanse of starry skies that graced the Iranian plateau as if the heavens themselves descended upon the earth. From flowers and trees to animals, from babbling brooks to tranquil lakes, and from sandy beaches to solid rocks and soil, I cherished every facet of nature. In fact, the unspoiled majesty of nature, untarnished by human interference, always provided me with a glimpse of the celestial bliss that awaits us in paradise.[113]

The foundation of his environmental philosophy is this awareness of the tight relationship between "nature" and "paradise." He came to the conclusion that:

> At its core, the ecological crisis stems from a profound spiritual crisis that afflicts modern humanity. The inner turmoil and darkness that plague the human soul often spill over onto the environment, resulting in the wanton destruction of its delicate balance and harmonious order.[114]

The Basic Features of Nasr's Thoughts

Perennial Philosophy, which asserts that all religions and philosophical systems share a fundamental unity, forms the basis of Nasr's philosophy. Nasr's metaphysics centers on the concept of Being, which he sees as the ultimate reality that transcends all phenomena. He believes that the material world is not the only reality, but rather is only a manifestation of the divine reality that lies beyond it. In his view, the universe is an interconnected whole that reflects divine attributes and principles. In terms of cosmology, Nasr sees the universe as a sacred and interconnected whole. He believes that human beings have a responsibility to live in harmony with nature and

[113] Seyyed Hossein Nasr, "An Intellectual Autobiography," in The Philosophy of Seyyed Hossein Nasr, ed. L.E. Hahn (2001), 28.

[114] "Reply to Giovanni Monastra," in The Philosophy of Seyyed Hossein Nasr, ed. L.E. Hahn (2001), 516.

to respect the natural order of the cosmos. In his view, nature is not simply a resource to be exploited for human purposes but rather a manifestation of the divine. Nasr's approach to Islam is deeply traditional, and he has written extensively on the importance of preserving and transmitting the teachings of the Qur'an and the Prophetic Tradition. He sees Islam as a comprehensive system of thought and practice that provides a framework for the cultivation of both individual and collective spiritual growth. In his view, Islam emphasizes the importance of living in harmony with nature and cultivating a sense of responsibility towards the environment.

Nasr's environmentalism is grounded in his metaphysical and religious beliefs. He argues that human beings have a duty to protect and preserve the natural world, which is a reflection of the divine reality. He believes that the environmental crisis is fundamentally a spiritual crisis, and that a true solution requires a transformation of human consciousness and behavior. In his view, environmentalism is not simply a matter of protecting the natural world, but rather of cultivating a deeper sense of reverence and respect for the divine reality that permeates all of creation. Nasr's philosophy, religion, and ecological ethics are deeply intertwined and reflect a holistic approach to understanding the world and our place within it. He sees the natural world as a reflection of the divine and believes that we have a responsibility to live in harmony with nature and to cultivate a sense of reverence and respect for the natural order of the cosmos.

Nasr on Perennial Philosophy, Metaphysics, and Cosmology

Nasr's philosophical outlook has been deeply shaped by the perennial school of thought, which was established in the 1920s by René Guénon (1886–1951) and championed by Frithjof Schuon (1907-98) throughout the 20th century. This particular school of perennial philosophy stands out from its historical counterparts due to its emphasis on adhering entirely to a specific religious tradition.[115] Consequently, it is not uncommon for adherents of the Guénonian tradition of perennial philosophy to refer to

---

115 Kenneth Oldmeadow, "Traditionalism: Religion in the Light of the Perennial Philosophy" (Sri Lanka Institute of Traditional Studies, 2000), 144.

themselves as Traditionalists. [116] For the purposes of this discussion, we shall utilize this term to describe those who fall under this category. Nasr's ideas and perspectives closely align with those who have been profoundly influenced by the teachings of Guénon and Schuon. These include luminaries such as William Stoddart (b. 1925), Titus Burckhardt(1908-84), Ananda Coomaraswamy (1877-1947), Martin Lings(1909-2005), Lord Northbourne (1896-1982), Marco Pallis (1895-1989), Hasan Gai Eaton(1921-2010), and Huston Smith (b. 1919). Comprehending the essential principles of perennial philosophy aids in comprehending Nasr's dual function as both an Islamic and Traditionalist philosopher. Nasr and Frithjof Schuon share almost identical viewpoints on the perennial philosophy. [117] This philosophy posits that every religion has certain metaphysical truths that are essentially the same among all religions. The Supreme Reality, which encompasses all possibilities and renders anything beyond it incomprehensible, is considered the most crucial and fundamental principle of reality, owing to its absolute and infinite nature.[118]

The second key tenet of the perennial philosophy is that the infinite Single Reality contains multiple entities, which implies the existence of a hierarchical model of being and awareness that originates from the same Supreme Entity.[119] While these ideas are present, in some form, in all genuine religious traditions, they are fully elucidated only in the interior dimension of religion. [120] The perennial philosophy posits that a religion can only be considered genuine if it has its foundation in a revelation from the Supreme

[116] Huston Smith, "Nasr's Defense of the Perennial Philosophy" in "The Philosophy of Seyyed Hossein Nasr, The Library of Living Philosophers, Volume XXVII, eds." Lewis Edwin Hahn, Randall E. Auxier, Lucian W. Stone, Jr. (Chicago and La Salle: Open Court Publishing Company, 2001), 141.

[117] Patrick Laude, "Seyyed Hossein Nasr in the Context of the Perennialist School," in "Beacon of Knowledge: Essays in Honor of Seyyed Hossein Nasr, ed. Mohammad Faghfoory" (Louisville, KY: Fons Vitae, 2003), 246.

[118] Seyyed Hossein Nasr, "Traditionalist Approach to Religion," in The Essential Seyyed Hossein Nasr, ed. William C. Chittick (Bloomington, IN: World Wisdom, 2007), 21.

[119] Nasr, "Traditionalist Approach to Religion," 21. Also see, Frithjof Schuon, "Survey of Metaphysics and Esotericism" (Bloomington, IN: World Wisdom, 2002), 19-20.

[120] Huston Smith, "Nasr's Defense of the Perennial Philosophy," in "The Life and Thought of S. H. Nasr, ed. Lewis Edwin Hahn" (Albany, NY: State University of New York Press, 2010), 148.

Entity. Thus, true religion cannot be created solely through human efforts.[121] According to the perennial philosophy, authentic religious traditions must rely on direction from the Supreme Entity. This is because only a religion grounded in such guidance can lead individuals to comprehend the universe and transcend various stages of existence and consciousness to reach that ultimate Source.[122] Thus, Hinduism, Buddhism, Judaism, Christianity, Islam, and any other tradition that adheres to the three essential principles outlined above are regarded as authentic religious traditions by the perennial philosophy.[123]

The presence of any religion suggests that humans have a purpose, which involves comprehending the universe and interpreting their experiences. This suggests that all things in the universe have significance and intention beyond just human understanding. The philosophy that endures over time suggests that there are three fundamental principles that support every authentic religion: the oneness of the highest principle, a hierarchical organization of existence, and the ultimate purpose or meaning of all elements within the cosmos. This indicates that every entity has significance beyond its practical value to humans. Islam's external and internal dimensions both exhibit certain principles that are evident, particularly in its metaphysics. These principles are closely aligned with those of perennial philosophy, which is why Nasr, a traditional Islamic philosopher and Traditionalist, subscribes to them. Nasr's perspective on the ecological crisis and its solution rests on these perennial ideas and their implications. When questioned about the veracity of these principles, Nasr and other Traditionalists argue that they derive from revelation and a mode of perception that surpasses reason. Nasr's epistemology recognizes the presence of another cognitive ability in humans, and this forms the crux of his philosophy.

Nasr argues that the contemporary usage of "intellect" and "intellectual" pertains solely to the analytical aspects of the mind, and has lost its original

121 Nasr, "Traditionalist Approach to Religion," in "The Essential Seyyed Hossein Nasr", 22-23.

122 Ibid., 21.

123 Seyyed Hossein Nasr, "Reply to Sallie King," in "The Philosophy of Seyyed Hossein Nasr, The Library of Living Philosophers, Volume XXVII, eds." Lewis Edwin Hahn, Randall E. Auxier, Lucian W. Stone, Jr. (Chicago and La Salle: Open Court Publishing Company, 2001), 226 and 229.

association with contemplation, as per Nasr's view.[124] The concept of Intelligence, a human cognitive ability that exceeds reason in terms of insight and comprehensiveness, is acknowledged by Nasr and other traditionalists. This faculty is used in the act of perçeption, known as "Intellection." In both Eastern and Western religious traditions, there is recognition of levels of awareness beyond the ordinary.[125] In both Eastern and Western traditional philosophies, a clear differentiation is made between reason and intellect. Plato, for instance, emphasizes that intellect, surpassing reason, can comprehend eternal and unalterable Truth.[126] A renowned Islamic theologian Abu Hamid al-Ghazzali (d.1111), claimed that there exists a "light that illuminates the heart" enabling one to perceive the truth of the Unity of God (tawhid) and that everything originates from a single source. Al-Ghazzali referred to this perceptual ability as the "prophetic power," which goes beyond the scope of ordinary reasoning.[127] [128]

In the Quran, the term *al-'aql* is used to refer to both reason and intelligence, according to Nasr. In Islamic philosophy and Sufism, there exists a distinct separation between these two ideas. The interconnection between them and the reliance of reason on Intellect are consistently recognized. [129] There exists a noteworthy distinction, in the Islamic peripatetic philosophy of al-Farabi (870-950) and Ibn Sina (980-1037), between conventional human reasoning and the rational capability that is stimulated by an intelligence surpassing human levels.[130]

---

124 Seyyed Hossein Nasr, "Science and Civilization in Islam, 2nd edition" (Cambridge, UK: The Islamic Texts Society, 2007), 24.

125 Martin Lings, "Ancient Beliefs and Modern Superstitions, 2nd edition" (London and Boston: Unwin Paperbacks, 1980), 27-30.

126 Plato, Timaeus 3:28.

127 Abu Hamid al-Ghazzali, Al-Ghazzali On "Trust and the Unity of God, trans. Muhammad Nur Abdus Salam" (Chicago: Great Books of the Islamic World, 2002), 22.

128 Abu Hamid al-Ghazzali, Al-Ghazzali's "Path to Sufism: His Deliverance from Error, trans. R.J. Mccarthy, S.J". (Louisville, KY: Fons Vitae, 2000), 61. "That light could also be seen as being directly from God" Who is "the Light of the heavens and the earth" and Who "guides to His Light whom He wills." See Qur'an 24:35.

129 Seyyed Hossein Nasr, "Islamic Philosophy from its Origin to the Present: Philosophy in the Land of Prophecy" (Chicago: University of Chicago Press, 2006), 94.

130 See Osman Bakar, "Classification of Knowledge in Islam: A Study in Islamic Philosophies of Science" (Cambridge, UK: Islamic Texts Society, 1998), 62.

The philosophers affiliated with the Ibn 'Arabi school, such as Mulla Sadra, as well as those who followed Shihab al-Din Suhrawardi in Illuminationist philosophy, make the difference between rational thinking and a perceptual ability to perceive spiritual reality.[131] Schuon describes the Intellect as the light of God within human beings, which Meister Eckhart characterizes as "something in the soul that is uncreated and uncreatable." [132] Hence, our comprehension of God is solely possible by means of the divine spark residing within us, rather than relying on our mere human capabilities. Furthermore, the Intellect is comparable to the Qur'an's internal revelation, which offers direct divine guidance.[133] The Intellect is an integral aspect of human nature and remains partly active in every individual. However, when affected by the constraints of personal awareness, the Intellect functions akin to reason, merely imparting peripheral knowledge through its analytical role in the mind. [134] [135] Thus, reason functions as a human-level manifestation of the Intellect, as it executes the analytical role of the mind. However, it remains unclear how the Intellect can function without being limited by human constraints.

To reach the Intellect or the eye of the heart, the Sufi way to connect with God involves a demanding procedure of cleansing the heart. According to Schuon, this demanding spiritual journey requires the cultivation of virtues such as self-effacement, generosity, and love of truth, which he defines as a conscious and constant effort towards perfection.[136] Nasr reinforces Schuon's perspective by linking it to a Qur'anic statement that confirms the notion of tawhid, or the oneness of God, being naturally recognized by a sound and well-balanced human intellect, which can only be obstructed by the corrupting force

---

[131] See Nasr, "Islamic Philosophy from its Origin to the Present, 99. For an account of Mulla Sadra's views on the hierarchy of the faculties of perception," see Badr al-Din Shirazi, "The Wisdom of the Throne: An Introduction to the Philosophy of Mulla Sadra, trans. James Morris" (Princeton: Princeton University Press, 1981), 132.

[132] Frithjof Schuon, "The Transcendent Unity of Religions, 2nd Quest edition" (Chicago: Theosophical Publishing House, 2005), xxix.

[133] See Schuon, "Survey of Metaphysics and Esotericism," 81-82. Also, see Nasr, "The Need for a Sacred Science," 10.

[134] Frithjof Schuon, "In the Wake of the Fall," in "Science and the Myth of Progress", ed. Mehrdad M. Zarandi (Bloomington, IN: World Wisdom, 2003), 24.

[135] Seyyed Hossein Nasr, "Contemporary Man between the Rim and the Axis," in "Science and the Myth of Progress, ed. Mehrdad M. Zandi" (Bloomington, IN: World Wisdom, 2003), 103.

[136] Schuon, "In the Wake of the Fall," 23-24.

of desires that blur its perception.[137] To transcend the constraints of human consciousness, it is possible to cultivate virtues aligned with the teachings of a particular religious tradition that is grounded in Revelation. By doing so, the Intellect gains greater freedom to perceive the world with greater clarity over time.[138] [139] Another way of expressing this is that knowledge acquired through the Intellect is more "concrete" than knowledge gained through reason, which is inherently discursive and "abstract," lacking the depth of knowledge obtained at the core of one's being or within the depths of the heart.[140] The different stages of developing virtues or purifying oneself correspond to a hierarchical progression in the way the Intellect operates, which is occasionally known as the hierarchy of Intellects.[141]

Nasr and other adherents of Traditionalism rely on enduring principles and their ramifications present in religious revelations, as well as their own wise judgments derived from Intellection. Those who dispute the authenticity of Intellection can only argue that the evidence of Intellection is apparent in the manifestations of Intellection itself. Attempting to demonstrate the authenticity of Intellect to every individual is akin to proving the legitimacy of a particular religion to each and every person, which is an impossible task.[142] According to Schuon, every truth that liberates contains an inherent element that is self-evident. Although this element may not always be apparent to us, we understand it better as we recognize an implicit essence of our own being within the demonstration.[143] It is this understanding that inspires faith."[144] Therefore, the crucial factor in recognizing the perennial

---

137 Nasr, "An Introduction to Islamic Cosmological Doctrines" (Albany, NY: SUNY Press, 1993), 7, chap. 3.4-3.4.1. For Qur'anic verses, see chapter 3.4-3.4.1.

138 Schuon, "The Transcendent Unity of Religions" (Wheaton, IL: Theosophical Publishing House, 1984), 151.

139 *Nasr, "The Need for a Sacred Science" (Albany, NY: SUNY Press, 1993), 54.*

140 Nasr, "Science and Civilization in Islam" (Cambridge, MA: Harvard University Press, 1968), 24.

141 Seyyed Hossein Nasr, "Reply to Huston Smith," in "The Philosophy of Seyyed Hossein Nasr", ed. Lewis Edwin Hahn, Randall E. Auxier, and Lucian W. Stone, Jr., "The Library of Living Philosophers, vol. XXVII" (Chicago and La Salle: Open Court Publishing Company, 2001), 160.

142 Frithjof Schuon, "Logic and Transcendence, trans. Mark Perry, Jean-Pierre Lafouge, and James Cutsinger, ed. James Cutsinger" (Bloomington, IN: World Wisdom, 2009), 27.

143 Schuon, "Logic and Transcendence," 27.

144 Ibid.

principles is not just the intellectual discovery of a shared metaphysical identity across various religious traditions, but rather the direct awareness of the Intellect. Nasr adds that the universality of these principles serves to reaffirm the role of the Intellect.[145]

To achieve a clear comprehension of Nasr's philosophy, it is crucial to elucidate the distinction between the Traditionalist interpretation of ideas like metaphysics and cosmology and their contemporary philosophical or scientific interpretations. This will prevent any confusion and enable a genuine understanding. Nasr and other Traditionalists posit that metaphysics is not just a branch of philosophy, but the very science of Reality – the fundamental and comprehensive knowledge that precedes and encompasses all other disciplines."[146] Nasr argues that the Traditionalist understanding of metaphysics is equivalent to the interpretations of *m'arifah, jñana,* and other similar terms used in other religious traditions.[147] The concept of metaphysics is central to all religions, and it is defined by these traditional terms as the study of comprehending God at the most profound level. In the Hindu, Islamic, and Christian traditions, the most prominent metaphysicians were Meister Eckhart (1260-1327), Ibn Arabi (1165-1240), and, Shankara (788-821), respectively.

The explanation provided above posits that the perennial philosophy offers a universal account of traditional metaphysics. Nasr contends that Islamic metaphysics shares fundamental similarities with perennial philosophy. A Traditionalist adheres to the established metaphysics of their religion but also employs the perennial philosophy to comprehend other faiths. The principal source of metaphysical knowledge or certainty is the Intellect, which is a faculty of perception that unveils the authentic nature of things, transcending their superficial appearances. This is because the metaphysics of all religions is centered around the understanding of God or the ultimate truth.[148]

---

145 Seyyed Hossein Nasr, "Knowledge and the Sacred" (Albany: State University of New York, 1989), 132.

146 Ibid.

147 Seyyed Hossein Nasr, "Knowledge and the Sacred" (Albany: State University of New York Press, 1989), 132,).

148 Nasr, "Need for a Sacred Science" (Albany: State University of New York Press, 1993), 54, accessed March 23, 2023, https://www.jstor.org/stable/10.7864/j.ctt12805fd.

The contemporary philosophical interpretation of the term stands in sharp contradiction to the Traditionalist interpretation of metaphysics. Metaphysics is "a subfield of philosophy," according to Harvard University modern philosopher Peter Godfrey-Smith.[149] As for the subject metaphysics deals with, Godfrey-Smith says that "typical inquiries in this field involve the essence of causality, the existence of the "external world"... The phrase is occasionally viewed as pertaining to an examination that surpasses what can be tackled via scientific means. Framed in this manner, a substantial number of individuals consider metaphysics to be a misguided pursuit."[150] The present definition of metaphysics confines itself to the physical and rational universe.

The traditional and contemporary scientific understandings of cosmology are clearly distinct from one another. The Traditionalist interpretation of cosmology, like their understanding of metaphysics, is consistent with how cosmology has been interpreted by different religious traditions that aim to describe the cosmos as a whole, including both material and non-material realities, including God or the Highest Reality.[151] They show a hierarchical system of existence, authority, and existence in which God is the source and source of all material and nonmaterial beings and things at all levels of the universe. Metaphysics has traditionally been defined as the study of the Real or the study of how everything in the cosmos relates to the Ultimate Real, or God.[152] Therefore, classical cosmology is a representation of the universe as seen through metaphysical ideals. Titus Burckhardt defined traditional cosmology as the field of knowledge that seeks to comprehend the world in relation to its underlying cause, which he identified as Being.[153]

In traditional Christian or Islamic cosmology, the planet is situated at

---

149 Peter Godfrey-Smith, "Theory and Reality: An Introduction to the Philosophy of Science" (Chicago: The University of Chicago Press, 2003), 237.

150 Ibid. 237-38.

151 E. Edson and E. Savage-Smith, "Medieval Views of the Cosmos: Picturing the Universe in the Christian and Islamic Middle Ages" (Oxford: University of Oxford Press, 2004).

152 Ibn 'Ata'illah al-Iskandari, "The Book of Wisdom," chapter 1, verse 14, trans. Victor Danner (New York: Paulist Press, 1978), 49.

153 Titus Burckhardt, "Traditional Cosmology and Modern Science," in "Mirror of the Intellect: Essays on Traditional Science and Sacred Art," trans. and ed. William Stoddart (Cambridge, England: Quinta Essentia, 1987), 17.

the lowest level of a hierarchy of angelic beings with God occupying the highest level. However, when viewed metaphorically, every aspect of cosmology reflects a correlation between God and the universe, revealing the laws of unity and hierarchy that govern reality. Thus, the cosmos is a coherent whole that displays God's power, greatness, will, and wisdom at all levels, including the planet. Consequently, every element of the material world assumes a significant function. In contrast, contemporary cosmologies represent cosmographies that only focus on the material aspects of the universe based on current scientific discoveries and hypotheses. Hence, Nasr contends that no matter how much we comprehend the material facet of the universe, we cannot grasp its essence using the conventional approach of modern scientific cosmology.[154]

Traditional cosmology is undeniably a religious cosmology, and the term "religious worldview" is interchangeable with it. Within a single religious community, there may be several cosmologies, as different symbols may convey the same metaphysical connection between God and the universe. Similar to how different religious traditions have used distinct language and symbols to portray their cosmologies.[155] Consequently, when discussing a theological worldview, Traditionalists highlight fundamental metaphysical concepts. Thus, a religious worldview can be viewed as an expression of reality since, in every religion, God embodies what is ultimately the most real.

Therefore, for a Traditionalist, any theological worldview essentially relates to the idea that the universe has meaning or a purpose and that reality as a whole is defined by unity, hierarchy, and significance. Nasr contends that in addressing the environmental catastrophe, one must consider cosmology or worldview, as human decisions and actions are never made in a metaphysical vacuum, but are shaped by one's understanding of reality both within and beyond oneself. [156] Gestalt psychologists support this view,

---

154 Seyyed Hossein Nasr, "Man and Nature" (Chicago: University of Chicago Press, 2014), 22-23.

155 Martin Lings, "The Past in the Light of the Present," in The Underlying Religion: An Introduction to the Perennial Philosophy, eds. Martin Lings and Clinton Minnaar (Bloomington, IN: World Wisdom, 2007), 35-54.

156 Seyyed Hossein Nasr, "Reading the Cosmic Qur'an," lecture delivered to the Muslim Students' Association (MSA) at the George Mason University (USA), 27 March 2008.

positing that one's worldview influences their perception of the world.[157] For instance, we do not jump off a mountain because we understand that such an action would result in severe harm. Similarly, when someone prays, it is because they hold a belief in a God who exists in the universe and responds to their prayers. Nasr argues that the transition from a religiously-based worldview that dominated prior to the Scientific Revolution to a contemporary science-based worldview has resulted in human behaviors that have played a role in exacerbating the ecological crisis.

Nasr and Traditional Islam and Environmentalism

Nasr suggests that a retrospective analysis of the Islamic world from two centuries ago would reveal that all facets of the tradition, ranging from the mystical musings of a Sufi mystic to the legal edicts of a scholar, from the rigorous theological convictions of a Hanbalite scholar from Damascus to the wavering declarations of an extreme form of Shi'ism, have varying degrees of foundation in the Islamic tradition. This tradition is comparable to a solitary tree with Divine roots in the Qur'an and hadith, while its trunk and branches constitute the body of tradition that has thrived for over fourteen centuries across almost every inhabited region of the globe.[158] It is possible to argue that the Qur'an and hadith played a crucial role as the fundamental sources for seemingly inconsistent declarations, specifically concerning indispensable realities like the unity of God (tawhid), the structure of existence, and the beginning and end of the universe within God. According to Nasr, the definition of tradition is mainly established by its fundamental principles:

> The term "tradition" encompasses more than just routine behaviors or traditions, and it does not solely imply the effortless transfer of concepts and themes from one generation to the next. Instead, tradition represents a collection of principles that originated from the divine realm and are intrinsically linked to a specific manifes-

[157] Huston Smith, "Nasr's Defense of the Perennial Philosophy," in "The Essential Seyyed Hossein Nasr", ed. William C. Chittick (Bloomington, IN: World Wisdom, 2007), 151.

[158] Seyyed Hossein Nasr, "Traditional Islam in the Modern World" (London and New York: Kegan Paul International, 1994), 11

> tation of the Divine. As a result, these principles are implemented and employed in diverse circumstances and periods within a given society.[159]

According to Nasr, even though different religious groups held distinct beliefs and practices, they shared most of the Prophet Muhammad's practices and fundamental values. This differed from the present situation, where many Muslims' perspectives are influenced by what he terms "modern." Nasr defines "modern" as that which is separated from the Transcendent and unchanging principles revealed through revelation that govern all things. [160] Gerard Kelly has stated that modernism's core concepts rely on the dominance of reason, articulated through scientific investigation, and leading to technological progress. Nasr claims that the origins of the modern can be traced back to the scientific thinking of the European Renaissance, which reached its peak with modern science, culminating in the Enlightenment, scientism, and Industrial Revolution. Therefore, Nasr perceives modernism as a set of concepts that depend directly or indirectly on the rationality and denial of the transcendent aspect of contemporary science. [161] Nasr's perspective on modernism is reflective of his overall view of Islamic civilization's decline. He believes that the separation of reason from revelation and the rejection of the transcendent aspect of life in the modern world is the root cause of this decline. [162] Therefore, he argues for a revival of traditional Islamic scholarship and a reintegration of reason and revelation. Nasr's call for a return to traditional Islamic scholarship and a balance between reason and revelation has influenced Islamic thought and discourse, both in the West and the Muslim world.

According to Nasr, the presence of modernism and contemporary "fundamentalist" groups, which espouse beliefs that stem from modernism,

---

159 "Subsequently", in "Islam and the Plight of Modern Man," 2nd edition (Chicago: ABC International, 2001), 73.

160 Nasr, Seyyed Hossein. "Traditional Islam in the Modern World" (London: Kegan Paul International, 1990), 98.

161 Gerard Kelly, "Get a Grip on the Future without Losing Your Hold on the Past" (London: Monarch Books,1999), 133 quoted in Haifaa Jawad, "Seyyed Hossein Nasr and The Study of Religion in Contemporary Society," The American Journal of Islamic Social Sciences 22, no. 2 (2005): 66, n.18.

162 Bennett, David. "Unthinking Faith and Enlightenment" (London: Routledge, 1987), 7.

has given rise to a necessity to differentiate tradition from the impacts of modernity on society. Nasr argues that "traditional" Islam does not denote a novel form of Islam, but rather a reading of Islam that is free from the influence of modern scientific viewpoints or contemporary ideologies that are grounded in that paradigm. Nasr's definition of traditional Islam emphasizes the importance of Islam's inner dimension, specifically Sufism, which distinguishes it from other modern versions of Islam that deny this aspect. However, Nasr does not claim that all members of the community should follow his teaching.[163] Nasr's interpretation of traditional Islam is consistent with the views of major authorities in the Sunni and Shi'ite worlds, such as al-Ghazzali (d.1111) and Shaykh Baha' al-Din 'Amili (1546-1621), respectively. These scholars also stressed the importance of both the exterior and interior dimensions of Islam. By emphasizing the inner dimension of Islam, Nasr is not rejecting the importance of the outer aspects, but rather he seeks to restore balance to the religion. In doing so, he aims to offer an alternative to the extremes of modernism and fundamentalism that have become pervasive in contemporary Islamic discourse.[164]

Nasr underlines the significant contrast between the traditional stance and other contemporary movements, arising from the former's recognition of sapiential readings of the Qur'an and hadith, support for Islamic art, focus on moral values in economics, non-coercive governance, and refusal of the scientific viewpoint in preference for a spiritual/religious comprehension of the natural world. Nasr contends that Sufism and its associated metaphysical knowledge are vital in traditional Islam to maintain a religious/metaphysical comprehension of nature and establish a basis for Islamic ethics, which differs from the difficulties brought about by modern scientific perceptions of nature.[165] Nasr acknowledges the traditional nature of his own viewpoint in this regard:

> Growing up, my education was a top priority in my family, with frequent discussions on cultural and religious topics and Sufi

---

[163] Seyyed Hossein Nasr, "Traditional Islam in the Modern World" (London: Kegan Paul International, 1990), 19.

[164] Joseph Lumbard, "The Decline of Knowledge and the Rise of Ideology in the Modern Islamic World," in Islam, Fundamentalism and Betrayal of Tradition, ed. Joseph E.B. Lumbard (Bloomington, IN: World Wisdom, 2004), 39-77.

[165] Seyyed Hossein Nasr, "Traditional Islam in the Modern World", 14-15, 18.

> poetry as common as the morning breeze. My parents were instrumental in my education from a young age, devoting countless hours to teaching me Quranic verses, Persian poetry, and sacred history, even when I was just a preschooler. Although Tehran was undergoing gradual modernization, we resided in an Islamic town that retained the character of the medieval era. I still recall vividly my childhood home, the narrow alleys that led to it, the small mosque, the religiously adorned water fountain (saqa-khanah) nearby, the pious and kind neighbors, the Quranic recitations, and the calls of the peddlers passing by. These memories provide me with a concrete experience of the pre-modern world that I would later rediscover through my intellectual pursuits.[166]

From his mid-twenties onward, Nasr's early traditional Islamic education was enriched by two significant experiences. At the age of twenty-four, he became a member of the 'Alawi branch of the Shadhili Sufi order, which had a profound impact on his spiritual and intellectual development. Over the next twenty years, he pursued his studies under the guidance of some of the most esteemed traditional teachers in Iran. Among these teachers were luminaries such as Sayyid Muhammad Kazim 'Assar, Sayyid Muhammad Saba'abi (1892-1981), who authored the Tafsir *al-mizan* ("The Commentary of the Balance"), Sayyid Abu'l-Hasan Qazwini, and Hadi Ha'iri, an eminent expert on Rumi and Sufism who became a mentor and father figure to Nasr. [167] He focused primarily on traditional philosophy and metaphysics with his first three mentors, delving deeply into the works of Ibn 'Arabi, Suhrawardi, and Mulla Sadra. With Hadi Ha'iri, he also studied the rich tradition of Sufi literature. After a prolonged period of studying Western philosophy, Nasr underwent a transformative shift in his intellectual and spiritual outlook. He began to explore Sufism in greater depth, discovering a profound resonance with its teachings and practices. From this point on, he dedicated himself to the study of Islamic scholars, enriching his understanding of the Islamic tradition and deepening his appreciation for its insights into the nature of reality and the human condition.

He expresses his wonder at the profundity of Islamic knowledge and his

---

166 Ibid 7

167 Nasr, "An Intellectual Autobiography," in "*The Philosophy of Seyyed Hossein Nasr*", 27.

persistent dedication to it ever since: "Following an extensive exploration that involved studying numerous schools of Western philosophy and science, I realized that the works of Sufi masters and Islamic philosophers deeply resonated with me. As I underwent my own personal rediscovery, which involved significant searching and suffering, Islamic wisdom became an incredibly vivid and powerful reality in my life. I consider myself fortunate to have been guided by the grace of Heaven toward the eternal Sophia, of which Islamic wisdom is one of the most fundamental and all-encompassing embodiments. Since that moment, I have been fully committed to the intellectual path that I continue to follow to this day."[168] It was not solely for scholarly purposes that he embraced Islam. While he was a faculty member at the University of Tehran, Nasr's time as a student there prompted Dr Mohammad Faghfoory to reflect on those years, saying, "During the 1960s, numerous students at the University of Tehran struggled to comprehend how a young Harvard-educated professor managed to adhere to the Islamic practices of praying five times a day, regularly fasting during Ramadan, and observing Thursday fasts almost every week."[169]

Given that some scholars have suggested Nasr holds a Shi'ite intellectual perspective, it is important to take into account his stance on the Sunni/Shi'ite divide in the Muslim world. However, it should be noted that Nasr's affiliation with a Sunni Sufi order for the majority of his life contradicts this assertion, despite having predominantly Shi'ite Iranian professors. Furthermore, both the Sunni and Shi'ite worlds share the same fundamental philosophical questions regarding the nature of reality. Nasr argues that the differences between Sunnism and Shi'ism are primarily formal and legal and that Sufism transcends these outward distinctions without rejecting them. Thus, it is crucial to consider Nasr's perspective on the Sunni/Shi'ite divide while acknowledging his Sufi affiliation and the shared philosophical foundations of both branches of Islam.[170] Nasr contends that nearly all Sufi orders trace their initiatic chain

---

[168] Seyyed Hossein Nasr, "In the Quest of the Eternal Sophia," in "The Complete Bibliography of the Works of Seyyed Hossein Nasr," edited by Mehdi Aminrazavi and Zailan Morris (Kuala Lumpur: International Institute of Islamic Thought and Civilization, 1994), 28-29.

[169] Mohammad H. Faghfoory, ed., "Introduction to The Beacon of Knowledge: Essays in Honor of Seyyed Hossein Nasr" (Louisville, KY: Fons Vitae, 2003), xxvii.

[170] Seyyed Hossein Nasr, "The Heart of Islam: Enduring Values for Humanity"

back to the first eight Shi'ite Imams, while Ibn 'Arabi's influence extends to both Sunni and Shi'ite communities. Therefore, the claim that Nasr has a pro-Shi'ite bias, according to Iraqi Sunni Islamic scholar Haifaa Jawad, lacks substance. It is also noteworthy that Nasr's work has influenced predominantly Sunni nations such as Indonesia, Malaysia, Pakistan, Turkey, and his homeland Iran.

Nasr's interpretation of traditional Islam has gained widespread support from prominent Islamic scholars. John Voll, a renowned historian of the Muslim world, notes that Nasr has consistently presented an Islamic viewpoint that aligns with the majority consensus within the framework of Islamic traditions. Since the 1970s, Nasr has been recognized as the most prominent representative of Muslim intellectuals who are well-versed in both traditional Islamic thought and modern issues.[171] William Chittick, a notable scholar of modern Islamic intellectual tradition, argues that Nasr's analysis of the contemporary relevance of Islamic ideas is firmly rooted in tradition. Although some of Nasr's critics may refuse to acknowledge this fact, Chittick maintains that Nasr's frequent references to Traditionalist scholars like Schuon do not diminish the validity of his views. According to Chittick, Nasr quotes contemporary authors not to support his arguments as a preacher would, but rather as a philosopher who has found modern writers to be the most articulate sources for his intellectual perspective.[172] Nasr's articulation of traditional Islam has garnered support from leading Islamic scholars. His consistent adherence to the majority consensus within the framework of Islamic traditions, combined with his knowledge of modern issues, has made him a highly respected representative of Muslim intellectuals. Despite some criticism, Nasr's use of contemporary sources does not invalidate the Islamic foundation of his ideas, according to Chittick. Instead, Nasr's approach as a philosopher seeks to provide clarity and coherence to his perspective on Islam in the modern world.

Nasr frequently cites notable scholars of Islamic tradition, such as Ibn Sina,

---

(New York: HarperCollins, 2002), 62.

171 William C. Chittick, "Science of the Cosmos, Science of the Soul: Pertinence of Islamic Cosmology in the Modern World" (Oxford: Oneworld Publications, 2007), 78.

172 Ibid

al-Ghazzali, Ibn 'Arabi, Mulla Sadra, and Rumi, in addition to the Qur'an and Hadith. According to Jawad, Nasr is a significant advocate of the traditional Islamic viewpoint. [173] Jane Smith, another respected Islamic scholar, agrees and believes that Nasr convincingly argues for the essential need to reclaim the traditional Islamic perspective.[174]

After reviewing multiple country reports on Environmentalism in the Muslim World, it is evident that comprehensive research has been conducted on Muslim environmentalism throughout the world.[175] From these findings, several conclusions can be drawn about the state of ecology in the Muslim world today. One of the main contributors to the environmental deterioration in Muslim nations is development activities, which is also a significant issue in the West. Several important Muslim nations have adopted environmental movements inspired by their secular counterparts in the West, attempting to follow similar strategies in promoting environmental awareness and "managing" natural resources using secular or technological means. However, this serves as a confirmation of the shortcomings of traditional Western environmentalism, as previously discussed. Ali Ahmad, a Nigerian Islamic legal expert, laments the methodologies and strategies of action used by the majority of Muslim environmentalists in Nigeria. He observes that their approaches and courses of action fully conform to the Western model, without taking into account an Islamic perspective that would adequately account for local sensitivities.[176] Even Saudi Arabia, the birthplace of Islam, has a National Commission for Wildlife Conservation and Development (NCWCD) that is operated as a secular organization. As anticipated, the *hima* system in Saudi Arabia has been gradually deteriorating over several years, with much of it being replaced by inappropriate economic development projects.[177]

Environmentalism faces opposition from industrial and commercial

[173] Jawad, "Seyyed Hossein Nasr and the Study of Religion," 49.

[174] Jane I. Smith, "Seyyed Hossein Nasr, Defender of the Sacred and Islamic Traditionalism," in "The Muslims of America," ed. Yvonne Y. Haddad (New York: Oxford University Press, 1991), 91.

[175] Richard C. Foltz, ed., "Environmentalism in the Muslim World" (New York: Nova Science Publishers, Inc., 2005).

[176] Ali Ahmad, "Nigeria," in "Environmentalism in the Muslim World", ed. Richard C. Foltz (New York: Nova Science Publishers, Inc., 2005), 81

[177] Khalid, "Applying Islamic Environmental Ethics," 107.

sectors similar to that in the West. Governments, except possibly Iran and Malaysia, are often at odds with environmental movements. Although the concept of "sustainable development," which aims to ensure future resource availability and is backed in part by Western nations, the UNDP, and international environmental organizations, is widely accepted, little is done on the ground to promote environmental sustainability. Typically, development activities are carried out without sufficient consideration of their environmental impact. Current government laws are lenient and poorly enforced, and there are several reasons for the lack of active involvement in environmental protection efforts.

The environmental crisis is a global threat to the planet's ecology and will make life on Earth increasingly challenging in the future. Many people are not aware of the severity of the crisis and mistakenly believe it is limited to certain Western countries or just about garbage in their neighborhoods. However, as previously discussed, development activities driven by the concept of material progress and scientific progressivism are the primary cause of environmental damage. Nasr sees his role as an intellectual who can influence change at the level of ideas rather than actions. Our discussion will therefore focus on the intellectual frameworks provided by Muslim modernists and some "fundamentalists" who support and legitimize scientific progressivism. We will see that these reformers, starting with Jamal al-Din al-Afghani (1838-1897), and later encouraged by other "fundamentalists," do not perceive a fundamental distinction between medieval Islamic science and contemporary science. As a result, they view contemporary science and technology as value-neutral and the primary engine for societal advancement. However, according to Nasr, the lack of distinction between conventional and modern technology, coupled with scientific progressivism and ignorance of associated secular cultural baggage, pose significant obstacles to a spiritual vision of nature that is necessary to revive love and care for nature. Nasr argues that relying solely on ethical solutions to the ecological crisis will not be sustainable in the long term if modern technology is allowed to continue unchecked. Nevertheless, ethical strategies can be effective in certain contexts, such as the Misali Ethics Project in Zanzibar led by Fazlun Khalid, Othman Llewellyn's Jabal Aja project in Saudi Arabia, tree-planting campaigns in Northern Nigeria, and Ayman Awhal's efforts to preserve forests in Malaysia and Indonesia, which are relatively isolated from larger financial activities.[178] [179] [180] [181] However, the environmental deterioration in Iran demonstrates that environmental degradation cannot be halted or reversed without sacrificing the collective drive towards technological and economic development. While the Iranian government has emphasized the environmental ethics of Islam, more needs to be done to address the root causes of the country's environmental problems. [182]

Nasr largely agrees with the majority of other well-known Muslim environmentalists that Islamic environmental ethics need to be revived and expanded in order to address the problem of current environmental deterioration. The individuals we are discussing are prominent environmentalists in the Islamic community. They include Mawil Izzi Dien, Fazlun Khaled, S. Nomanul Haq, Othman Abd-ar Rahman Llewellyn, Ibrahim Ozdemir, and several others. Their environmental beliefs are rooted in traditional Islamic teachings and can be readily traced back to the Qur'an or the hadith.[183] Since Nasr's publication of Introduction to Islamic Cosmological Doctrines in the early 1960s, many Islamic environmentalists have drawn inspiration from his explanation of the interrelationship between God, humans, and the natural universe. However, they differ from Nasr in their perspectives on modern technology and the potential role of Islam's inner component in addressing the environmental crisis. For instance, Fazlun Khalid, arguably the most prominent Islamic environmentalist, emphasizes Islamic ethics but acknowledges Nasr's contributions to the field and seeks his counsel.

---

[178] Eliza Barclay, "African Fisherman find way of conservation in the Koran," Christian Science Monitor, October 31, 2007, accessed August 20, 2010, http://www.internationalreportingproject.org/stories/detail/407/.

[179] Ibid., 106-107.

[180] Ali Ahmad, "Nigeria," in "Environmentalism in the Muslim World," ed. Richard C. Foltz (New York: Nova Science Publishers, Inc., 2005), 81-82.

[181] Ayman Ahwal, "Saving Rivers and Forests in Malaysia," in "Environmentalism in the Muslim World", ed. Richard C. Foltz (New York: Nova Science Publishers, Inc., 2005), 113-131.

[182] Richard C. Foltz, "Iran," in "Environmentalism in the Muslim World," ed. Richard C. Foltz (New York: Nova Science Publishers, Inc., 2005), 6.

[183] "Islam and Ecology: A Bestowed Trust," edited by Richard Foltz, Frederick Denny and Azizan Baharuddin (Cambridge: Harvard University Press, 2003), essays by Ibrahim Ozdemir, S. Nomanul Haq, Fazlun Khalid, Abdul Aziz Said and Nathan Funk. Harfiyah Abdel Haleem, ed., "Islam and the Environment" (London: Ta-Ha Publishers Ltd., 1998).

Nasr seeks to address the crisis by integrating both the outer and inner dimensions of Islam, including Shari'a, Sufism, conventional sciences, and later Islamic philosophy, all of which incorporate Sufi metaphysics and uphold the metaphysical principles of Islam. Therefore, we believe that Nasr's response to the environmental problem is more genuinely Islamic than those of other Islamic environmentalists.[184]

Many Islamic environmentalists consider the current scientific view as ignorant and an obstacle to achieving a sustainable solution to the environmental problem.[185] However, unlike Nasr, they have not extensively criticized modern technology, scientism, or scientific progressivism, nor have they provided comprehensive solutions to help Muslims tackle scientism and scientific progressivism in the long run. Some Islamic environmentalists share the scientific progressivism of modernist reformists, such as Mawil Izzi Dien. He believes that the West's civilization effectively utilizes natural phenomena for human necessities through empirical methodology and organization. He further suggests that substantial progress can be made by merging the Islamic concept of value with this utilitarian civilization, especially in the Muslim realm and beyond. This assertion is surprising given his understanding of the catastrophic social upheaval and environmental calamity caused by the Middle East's rapid industrialization over the previous few decades.[186]

Nasr's View Towards Lynn White Jr.'s Book, and Technological Solutions of Ecological Problem

The discourse on religion's role in human responses to the ecological crisis has been largely influenced by Lynn White Jr.'s influential article "The Historical Roots of our Ecological Crisis" and subsequent Christian critique. Therefore, it's crucial to compare Nasr's position with White's to

184 Fazlun Khalid, "Applying Islamic Environmental Ethics," 89.

185 Ibrahim Ozdemir, S. Nomanul Haq, Fazlun Khalid, Abdul Aziz Said and Nathan Funk in "*Islam and Ecology: A Bestowed Trust,*" eds. Richard C. Foltz, Frederick M. Denny, and Azizan Baharuddin (Cambridge: Harvard University Press, 2003).

186 Mawil Izzi Dien, "Islam and the Environment: Theory and Practice," in "Islam and Ecology: A Bestowed Trust", eds. Richard C. Foltz, Frederick M. Denny, and Azizan Baharuddin (Cambridge, MA: Harvard University Press, 2003), 110-114.

fully comprehend the differences in their perspectives.[187] [188] Published ten months after Nasr's Rockefeller Lectures in May 1966, White's book argued that the Biblical notion of humans having "dominion" over the planet was the cause of the environmental crisis.[189] While White echoed several themes from Nasr's lectures, he did not provide the philosophical justification that Nasr had. However, White did concur with Nasr on the importance of our attitude toward nature and the lack of scientific solutions to the environmental problem.[190] According to Nasr, the idea of dominance that exists in modern times did not originate from Biblical teachings. Instead, it emerged during the Renaissance, Scientific Revolution, and modern technological era due to the glorification of human reason. According to Nasr, since the environmental problem arose after the Industrial Revolution, which itself was reliant on the Scientific Revolution, it is reasonable to conclude that the environmental crisis did not exist prior to these events.[191] In contrast to White's argument against Christian theology, Nasr posited that the rise of science and technology leading to environmental devastation was not unique to Christian cultures, as other factors must have been involved. Nasr made a unique point in the discussion about the idea of dominating nature. He argued that it goes against man's role as the custodian and guardian of nature, a perspective that has not often been considered.[192] [193]

According to Peter Harrison's research, the Hebrew word "rada" used in Genesis, commonly translated as "have dominion," actually implies the concept of fair and peaceful governance.[194] [195] Lynn White's argument

---

[187] Lynn White, Jr., "The Historical Roots of Our Ecological Crisis," Science Vol 155 (10 March 1967): 1203.

[188] Willis Jenkins, "After Lynn White: Religious Ethics and Environmental Problems," Journal of Religious Ethics 37, no. 2 (2009): 283.

[189] White, "The Historical Roots," 1205.

[190] Ibid., 1205-06.

[191] See Chapter 1.5-1.5.1.

[192] Nasr, Seyyed Hossein. "Need for a Sacred Science." Albany: State University of New York Press, 1993, 137.

[193] Nasr, Seyyed Hossein. "Man and Nature: The Spiritual Crisis in Modern Man." Chicago: Kazi Publications, 1997, 19.

[194] Genesis 1:26; 1:28.

[195] Steffen, Lloyd H. quoted in Peter Harrison, "Subduing the Earth: Genesis 1, Early Modern Science, and the Exploitation of Nature," The Journal of Religion 79, no. 1 (1999): 88.

also overlooked the fact that human dominance over nature did not imply exploitation, as Genesis 2:15 indicates that humans were placed in the garden to cultivate and care for it.[196] Similarly, Seyyed Hossein Nasr and other Islamic environmentalists reject the interpretation of Qur'anic texts that suggest God made the natural world subservient to humans as a justification for environmental exploitation.[197] According to Nasr, this understanding is limited to individuals who continue to serve God and does not extend to those who assert their autonomy from God's wishes. Another verse in the Qur'an (22:37) states, "Thus have We made them subject unto you that ye may magnify Allah," referring to animals.[198]

Although White's argument against the Christian belief in human supremacy over nature has been criticized by many as oversimplified, his indictment of Christian scripture for environmental issues sparked a wave of objections from Christian academics.[199] Nasr's The Encounter of Man and Nature did not receive much notice in contrast to White's article, apart from a few subsequent works. For instance, Theodore Roszak's Where the Wasteland Ends, initially released in 1972, and Phillip Sherrard's The Rape of Man and Nature, both contend that the ecological crisis is an external manifestation of humanity's contemporary spiritual dilemma. Alternatively, as Nasr explains, those who are at peace with God are also at peace with His creation, including both nature and humanity.[200] [201] [202]

More than four decades after Nasr's Rockefeller Lectures, there is a growing emphasis among major environmentalists on the need for spiritual revival as the crucial foundation for any effective solution to the environmental

---

196 Quoted in Duncan Roper, "The earth as a 'garden' for all creatures: Lynn White forty years on," Stimulus 15, no. 4 (November 2007): 20.

197 Qur'an 14:32-33.

198 Nasr, Seyyed Hossein. "The Islamic View of the Universe." In "Islamic Culture," vol. I, Foundation of Islam, edited by Z. Ansari, manuscript p. 27. Paris: UNESCO, in press.

199 Oelschlaeger, Max. "Caring for Creation: An Ecumenical Approach to the Environmental Crisis." New Haven: Yale University Press, 1994, 22-27.

200 Theodore Roszak, "Where the Wasteland Ends: Politics and Transcendence in Postindustrial Society" (Berkeley, CA: Celestial Arts, 1989), 52.

201 Philip Sherrard, "The Rape of Man and Nature" (Ipswich, UK: Golgonooza Press, 1987), 88.

202 Seyyed Hossein Nasr, "Man and Nature" (Chicago: ABC International Group, 2017), 136.

problem. During this time, Nasr has expanded on his book in his work *Religion and the Order of Nature* and has presented an overview of a specifically Islamic approach to the crisis through various publications, speeches, and interviews over the past decade. In doing so, he has considered contemporary intellectual challenges and drawn on Islamic ethics, philosophy, metaphysics, and scientific traditions, for which he is widely regarded as one of the foremost authorities. One could argue that to fully comprehend Nasr's approach to addressing the environmental crisis, it is necessary to acknowledge his explorations of traditional Islamic philosophy, metaphysics, and scientific traditions. Nevertheless, it's worth highlighting that Nasr, along with other Western environmentalists scrutinized, doesn't reject conventional scientific solutions to environmental challenges. To dispel any notion to the contrary, it is vital to highlight Nasr's stance on the role of modern technology in addressing the crisis.

According to Nasr, a long-term technological solution to the environmental problem is not desirable for several reasons. One reason is that modern science and technology, which have contributed to the problem, cannot be relied upon to provide a solution. In fact, turning to technology as a solution may only reinforce the same mindset that caused the problem in the first place.[203] For Nasr, the most important reason for avoiding a technological solution is that it undermines the vision of unity in reality, which is a fundamental aspect of his worldview.

Secondly, according to Nasr, modern science is incapable of comprehending the intricate interactions between the various components of a real ecology, even at a material level. Nasr highlights that while it is easy to calculate the gravitational force between two isolated objects, the presence of a third object makes the calculation extremely difficult, and the presence of a fourth object makes it nearly impossible. Furthermore, Max Oelschlaeger observes that the theory of chaos, in conjunction with other scientific advancements, reveals nature as a dynamic system of such complexity that it is impossible to provide a conclusive scientific depiction and theoretical explanation. [204] Since modern science is unable to identify the actual relationships among the

[203] Seyyed Hossein Nasr, "Islam and the Plight of Modern Man," 2nd edition (Chicago: ABC International, 2001), 4, 17.

[204] Max Oelschlaeger, "Caring for Creation: An Ecumenical Approach to the Environmental Crisis," (New Haven: Yale University Press, 1994), 40.

numerous variables in nature, there is no basis for assuming that a technological solution, even in the distant future, is possible. Nasr does not oppose technological or economic solutions if they can slow down degradation, despite the fact that the environmental issue is currently in a critical stage.

> It is an undeniable truth that taking immediate practical steps, such as expanding public transportation and switching from petroleum to natural gas, is crucial. Any efforts in this direction are praiseworthy as they can provide us with additional time to tackle the problem. I endorse all immediate technological or economic solutions as they can help us adopt cleaner technologies and mitigate the pollution. Al Gore and others' assertions are largely accurate, but I do not believe that solely relying on those technologies will rescue us from this crisis. We need a transformation from within. We require an alternative approach to viewing ourselves, human life's purpose, and what brings us happiness. We must not rely solely on consumption as the path to happiness.[205]

The overall objective of Meadows et al., Sachs, and Speth are not fundamentally different from Nasr's stance. However, Nasr's perspective goes beyond theirs in his criticism of modern science and technology and the associated worldview. He advocates for an alternative science that aligns with Islamic metaphysics, philosophy, and religion. To understand how Nasr's arguments relate to the environmental problem for people of different religions globally, it is necessary to briefly discuss the perennial philosophy and its perspective on conventional epistemology, metaphysics, and cosmology in various religious traditions.

---

205 Seyyed Hossein Nasr, "On the Environmental Crisis," interview by Muzaffar Iqbal, in "Islam, Science, Muslims, and Technology: Seyyed Hossein Nasr in Conversation with Muzaffar Iqbal" (Alberta, Canada: Al-Qalam Publishing, 2007), 128-129, 131.

Ecological ethics of John B. Cobb, Jr. and Seyyed Hossein Nasr

John B. Cobb, Jr.[206] and Seyyed Hossein Nasr are both notable scholars and thinkers in the fields of religion, philosophy, and ecology. John B. Cobb, Jr. is an American theologian, philosopher, and environmentalist. He is known for his work in process theology, a philosophical and theological approach that emphasizes the dynamic and interrelated nature of reality. Cobb has also been a prominent voice in the environmental movement, arguing that environmental problems are fundamentally linked to deeper cultural and spiritual issues.

Cobb and Nasr, two scholars who have delved deeply into eco-religious ethics, have presented their proposals for achieving environmental sustainability. Their works are widely recognized as influential contributions to this field from a religious perspective. Both scholars connect eco-religious teachings to current environmental discourse, and they reinterpret religious moral teachings related to the environment to achieve this aim. However, they use different approaches in their interpretation. Until 1969, Cobb had not been willing to shift his focus towards environmental issues. However, upon reading the influential works of Paul Ehrlich and Lynn White on the escalating ecological crisis, Cobb developed an interest in eco-theology.[207] [208] His aim was to demonstrate how the Christian faith could be utilized to reform human attitudes towards the natural environment. As a critic of the historical role of the Christian Church, he endeavors to reconstruct Christian teachings regarding nature and the environment. While acknowledging that the medieval Christian Church may have indirectly contributed to environmental degradation by endorsing a human-centered approach to the natural world, Cobb asserts that the present ecological crisis is largely caused by modern science and technology.[209] According to Cobb, the natural environment has suffered damage from both natural

---

206 For his life and works, see John B. Cobb, Jr., *"Theological Reminiscences"* (Claremont, California: Process Century Press, 2014). For his theological biography, see David Ray Griffin and Joseph C. Hough, *"Theology and the University: Essays in Honor of John B. Cobb"* (Albany: State University of New York, 1991).

207 White, "The Historical Roots of Our Ecologic Crisis."

208 Paul Ehrlich, *"The Population Bomb"* (New York: Ballantine Books, 1968).

209 Cobb, *"Sustainability,"* 1; Cobb, *Sustaining the Common Good*, vii.

and artificial sources over the centuries, with human civilization playing a significant role in this destruction. The European Renaissance, humanism, and secular perspectives on nature have worsened the situation. With modern scientific knowledge convincing humans to exploit nature for their own comfort and convenience, and with a mechanistic view of nature and the aid of advanced technology, humans have been causing unprecedented harm to the environment since the Industrial Revolution.[210]

As previously mentioned, Cobb expressed disappointment with the Christian Church's role in preserving the natural environment. He attributes this to the Church and Christian theologians misinterpreting biblical verses to justify exploiting the natural world. According to Cobb, the Bible not only depicts human superiority over non-human animals and plants but also designates humans as caretakers of God's creation. Cobb finds fault with the Church and theologians for consistently prioritizing human superiority and neglecting our responsibility as caretakers of the natural world. While modern scientists and philosophers were advocating for conquering the natural world through technology, the Christian Church and theologians were attempting to justify this view as a biblical prophecy, rather than emphasizing humanity's role as stewards of the Earth. Cobb contends that these theological interpretations were one-sided because Christian scholars placed too much importance on human interests, which he believes is an incorrect way of interpreting the Bible. By offering constructive criticism of the Church and theologians, Cobb takes on the challenge of reinterpreting biblical verses to reconstruct Christian moral teachings to promote ecological justice. His knowledge of process philosophy and process theology helps him with this task.

Nasr, much like Cobb, contends that modern science has been the sole criterion for knowledge since the emergence of the European Renaissance and humanism. According to him, spiritual knowledge has been excluded from the realm of modern knowledge. This kind of knowledge is supported by a secular ideology that strips nature of its sacredness. After prolonged debates and struggles between science and religion in Europe, the scientific revolution gained momentum and ultimately prevailed over theology. As a result, scientism became an unquestionable authority. Even religious

---

[210] Cobb, "*Sustainability*," 93.

institutions, particularly the Christian Church, could not challenge scientific theories and instead embraced the discoveries of scientists as proof of human superiority over the natural world, as predicted in biblical prophecy. Nasr adds that while the Christian Church initially opposed modern science, it eventually embraced it and even endorsed it. However, like Cobb, he criticizes the Church for not speaking out against the negative consequences of modern science. Nonetheless, Nasr believes that Christianity, as a significant faith tradition, can address the ecological crisis. He proposes the development of a new Christian theology, specifically eco-theology, to tackle present environmental problems from a Christian perspective. Nasr also incorporates the world's religious traditions into the discourse on the ecological crisis, although he tends to favor Eastern traditions over Western ones.[211]

Nasr believes that traditional ideas of nature hold significant value in achieving environmental justice. This is why he identifies as a traditionalist rather than a modernist or reformist. He argues that traditional ways of understanding hold the key to environmental sustainability, and all traditions are valuable treasures of the world that can be used to address ecological problems. Consequently, he advocates for the retrieval of traditional knowledge rather than constructing new ways of understanding nature. Although Nasr primarily focuses on Islamic faith traditions and spirituality, he acknowledges and respects the importance of other religious traditions.

Both scholars agree that modern science and technology are responsible for the current ecological crisis.[212] They suggest that religions have not played a positive role based on their moral foundations in nature. However, they believe that religious moral teachings can motivate human behavior towards treating the environment and non-human animals justly. To achieve environmental sustainability, they propose revitalizing such moral guidelines.[213] Cobb and Nasr attempt to connect their respective religions with the current ecological degradations to foster sympathetic consciousness for the natural world. While Cobb is critical of the role of

---

[211] Seyyed Hossein Nasr, "The Spiritual and Religious Dimensions of the Environmental Crisis," *"The Ecologist"* 30 (2000).

[212] Cobb, *"Is It Too Late?"* 33; Seyyed Hossein Nasr and Muzaffar Iqbal, "The Islamic Perspective on the Environmental Crisis," *Islam & Science* 5, no.1 (2007): 76.

[213] Cobb, *"Sustainability,"* 2.

the Christian Church and theologians, Nasr is not as critical of the Islamic traditional understanding of the natural world. Cobb's approach is more reformative and reconstructive, while Nasr's approach is more revival and explorative. Both scholars suggest reinterpreting the theology of nature to address current environmental problems from a faith perspective. Their initiatives have attracted theologians and researchers to contribute further to the environmental issue in theology and religion.

Cobb and Nasr consistently connect the natural world to living entities and contend that the ecological crisis arose when nature was perceived as lifeless through a mechanistic lens. They strongly condemn modern science and its purportedly secular scientific outlook for affirming nature's lifelessness and advancing a mechanistic perspective rooted in it. If nature appears lifeless to modern humans, they are unable to develop a compassionate attitude towards it. It is illogical to expect humans to hold intrinsic value in lifeless natural objects. Consequently, with this endorsement from modern science, humans are automatically inclined to exploit the natural world like a production machine to meet their limitless wants and needs. This makes it effortless for modern humans to treat nature irresponsibly since it is deemed dead and solely existent for their benefit; as an inanimate substance, nature is not deemed worthy of human attention. Cobb and Nasr argue that if nature were not perceived as dead, humans might not exhibit such reckless behavior towards the natural world; conversely, if it were viewed as a living entity, human attitudes towards nature could differ from what is presently observed.[214] [215]

According to Cobb and Nasr, the primary cause of the current ecological crisis is humanity's perception of nature, specifically the belief that nature is dead or lifeless. They argue that unless this perception is fundamentally altered to one that recognizes nature as alive, humans will not be motivated to treat the natural world with care and respect. To address this issue, both Cobb and Nasr have developed their own approaches to rectifying humanity's perception of nature, particularly the mechanistic view.[216] Cobb's

---

214 Cobb, *"Is It Too Late"?*, 25-28; Nasr, *"The Need for a Sacred Science"*, 119-125.

215 Charles Birch and John B. Cobb, Jr., *"The Liberation of Life: From the Cell to the Community"* (Cambridge:
Cambridge University Press, 1981), 263; Nasr, *"Man, and Nature"*, 14.

216 Birch and Cobb, *"The Liberation of Life,"* 17, 42, 123, 240.

ecological model is commendable as it portrays the natural world as an organism in which everything is interconnected and interdependent, rather than a machine or lifeless entity. He argues that modern biological science does not support the mechanistic view of nature. On the other hand, Nasr takes a metaphysical approach to demonstrate that although the external appearance of the natural world may seem lifeless, it is, in fact, alive. By referencing holy scriptures and the views of mystics, he aims to persuade humans that the mechanistic view is incorrect.[217] Therefore, Cobb and Nasr strive to represent nature through a metaphor for life, although their approaches differ. Cobb connects this metaphor to biological and ecological science, while Nasr links it to a spiritual and metaphysical understanding.

Cobb acknowledges the hierarchy of value and bio-pyramid that place humans above other beings, but he does not reject the intrinsic value of non-human entities in the environment. He uses biblical interpretations and scientific evidence to support his belief that other non-human beings have inherent value, citing Genesis 2:15. In contrast, he criticizes Christian scholars for misinterpreting Genesis 1:28 to deny the intrinsic value of other entities. Instead, Cobb suggests interpreting both Genesis 1:28 and 2:15 for a comprehensive understanding of the relationship between humans and non-human components of the natural world. According to Cobb, everything in creation has its own value, but not all values are equal. He argues that the Bible does not support extreme anthropocentric values, though it does give humans special dignity over other creatures.[218]

According to Cobb, the Bible has a God-centric approach rather than an anthropocentric approach. He believes that Christian scholars and theologians have mistakenly transformed this theocentric approach into an anthropocentric one over time. This is arguably a fault of Christian institutions and theologians, rather than Christianity itself. Cobb interprets Christianity from an ecological perspective and considers the intrinsic value of all biotic organisms, which aligns with a bio-centric approach. Additionally, his views on the relationship between living forms and their environment reflect an eco-centric approach, while his discussions of the hierarchy of values and biotic pyramid pertain to an anthropocentric approach. Thus,

---

[217] Nasr, *"Religion and the Order of Nature"*, 24, 271.

[218] Cobb, *"Is It Too Late?"* 117; Birch and Cobb, *"The Liberation of Life,"* 205.

Cobb's eco-theological understanding combines these three mainstream value approaches of environmental ethics in a significant relationship with a bio-centric value approach.[219]

While Cobb recognizes the value of the earth and the Gaia hypothesis as counter-movements to economism, he is critical of these approaches because of their potential to replace God. As humans respond to God, he believes that we should not accept any ideology that neglects God's role. Cobb opposes the idea of unnecessary disputes and unexpected debates between different environmental movements such as deep ecologists, social ecologists, eco-feminists, sustainable agriculturalists, and the green movement. He advocates for combining their efforts to tackle the unprecedented ecological crisis. Cobb does not rely solely on traditional stewardship concepts and present environmental ethics as he finds them insufficient to care for the environment. He has developed various frameworks, including ecological asceticism, ecological model of life, and ecological model of development, to address the current environmental crisis. Therefore, Cobb's eco-theological approach is inclusive of certain points, critical to some extent, and pragmatic in terms of implementation.[220]

Similar to Cobb, Nasr acknowledges the existence of hierarchical relationships among the components of the environment. According to him, such relationships are universal and are observed with slight variations in all religious and cultural traditions. Nasr's God-Human-Earth hierarchical relationship model is not associated with an anthropocentric value approach to environmental ethics. While an anthropocentric approach places humans at the center, Nasr's model situates humans between God and nature, recognizing that humans are part of the natural world and are responsible for its care on behalf of God. Although humans hold a prestigious status among God's creatures, they are not given unlimited permission to exploit the natural world for their own desires. Nasr strongly criticizes the anthropocentric value approach to the environment, which he considers a product of the modern scientific worldview. He argues that due to the secular worldview, modern humans have first separated God from humans and then humans from nature, leaving only humans as the central

---

219 Ibid.

220 Cobb, *Sustaining the Common Good*, 40; Cobb, "*The Earthist Challenge to Economism*," 7, 179.

focus, while God and nature have no value in this perspective.[221]

Moreover, Nasr criticizes the secular philosophers and scientists who try to depict nature as lifeless and mechanical, existing only to serve human needs. He sees this as a consequence of the modern scientific worldview and a radical anthropocentric approach to environmental ethics. Instead, he proposes the concept of "anthropocosmism," which emphasizes the interconnectedness of heaven, earth, and humans in a proper relationship. In traditional societies, people held this view and respected nature with love and affection.[222] However, Nasr observes that modern humans have replaced this view with a more anthropocentric one, leading to a lack of respect for the natural world. He believes that unless people change their perception of nature, they cannot solve the current environmental crisis. Therefore, he suggests rediscovering the sacred quality of nature, which has been lost in modern times. By recognizing the aliveness of nature and reconnecting with its sacredness, humans can work towards environmental sustainability. Nasr advocates for following traditional knowledge and spiritual values, which hold nature in high esteem, to address the environmental crisis.[223]

Nasr's views on nature align closely with both bio-centric and eco-centric value approaches to environmental ethics, but there are notable differences between his model and these approaches. While a radical bio-centric approach reduces humans to the same level as other non-human entities in the environment, Nasr maintains the special dignity of humans and never considers them on par with other animals. Instead, he recognizes the inherent value of all living organisms, irrespective of their relationship to humans. Similarly, while eco-centric approaches devalue the position of humans, Nasr emphasizes the intimate relationship between humans and God's creation, urging people not to harm this relationship. Therefore, while Nasr's views are related to both bio-centric and eco-centric approaches, they cannot fully comply with either approach due to the differences in the treatment of human beings.[224]

---

221 Nasr, *"The Need for a Sacred Science,"* 136, 140; Nasr, *"Religion and the Order of Nature"*, 161.

222 Ibid.

223 Nasr, *"Man, and Nature,"* 48.

224 Ibid.

Nasr disagrees with the earthism and Gaia hypothesis perspectives on the environment, as these movements replace God with the earth. Similarly, the hierarchical relationship model of God-Human-Earth proposed by Nasr does not align with other radical environmental movements such as deep ecology and social ecology. Deep ecologists believe that the environmental crisis is due to humans' perceived superiority over other living entities and suggest reducing humans' special status to achieve environmental justice. Social ecologists argue that environmental sustainability requires a complete overhaul of social structures. Eco-feminists advocate for eliminating the patriarchal social system to ensure environmental preservation. Nasr sees these movements as focused on the cause rather than recognizing the unique dignity of humans as God's representatives on earth. They lack a metaphysical understanding of the universe and do not acknowledge the potential qualities of humans that could lead to environmental sustainability.

Nasr recognizes the value of sustainable agriculture and the green movement but considers them inadequate to address the magnitude of the current environmental crisis. In Nasr's view, the existing environmental ethics fall short as they fail to evoke a deep emotional connection to nature within humans. He believes that the crisis is primarily a spiritual one and cannot be resolved without addressing this underlying cause. Nasr urges people to focus on this fundamental issue and reconnect with nature at a deeper level. Rather than adhering to a specific value-based approach to environmental ethics, Nasr emphasizes the traditional understanding of nature and the importance of a spiritual connection with it, which he considers the most effective way to reverse the current situation.[225]

Cobb and Nasr offer their own perspective on environmental sustainability, critiquing a rigid anthropocentric value approach to environmental ethics. While their views appear to align with bio-centric and eco-centric approaches, they do not fully conform to either. They do not support radical environmental movements like earthism or the Gaia hypothesis, nor do they reduce the special status of humans to other non-human entities as seen in the deep ecological and green movement. According to Cobb and Nasr, current environmental ethics is inadequate to address the unprecedented ecological crisis, which they see as a spiritual crisis as

[225] Nasr, "The Spiritual and Religious Dimensions of the Environmental Crisis."

well. They believe that a solution requires a sound spirituality, which is ignored in the secular scientific worldview that they both criticize. They emphasize the importance of heartfelt responsibility in the human mind towards the natural world and suggest that humans take sincere care of the environment as responsible creatures of God on earth. They focus on the responsible attitudes of humans towards non-human entities of the environment as sharing partners on the same earth, urging people to modify their present behaviors towards the natural world. Cobb's ecological model of life and Nasr's God-Human-Earth relationship model, though different from each other, appear unique and distinct from existing value approaches to environmental ethics.

Chapter Summary

Seyyed Hossein Nasr is a highly respected Islamic intellectual and environmental philosopher who has made significant contributions to the fields of philosophy, religion, and environmentalism. His upbringing in an academically accomplished family, his education in the US, and his extensive travels throughout the world have influenced his work greatly. Nasr's writings reflect his Islamic faith and his knowledge of Western philosophy and science, offering a unique perspective on the relationship between religion, science, and the environment. His works aimed at both academic and general audiences, have been translated into many languages and have earned him a distinguished reputation as a major intellect of our time. Nasr's belief in the restoration of religiously oriented lifestyles as an alternative solution to technology-driven modern lifestyles is a significant aspect of his environmental philosophy, and his call for a change in the scientific worldview, which he argues has been the root of the environmental crisis, is a fundamental concept in his philosophy. Nasr's vast body of work provides valuable insights into the nature of the universe, humanity's place in it, and the role of religion and spirituality in the modern world.

In this chapter, we learned about Nasr's unique perspective on these topics, which is influenced by his Islamic faith and knowledge of Western philosophy and science. Nasr believes that the restoration of religiously oriented lifestyles is a key solution to the ecological crisis caused by the contemporary scientific worldview.

In chapter two, we see how Nasr's philosophy informs his belief that the contemporary scientific worldview is responsible for the environmental problem. Nasr argues that human actions based on this worldview, which prioritizes materialistic objectives and ambitions, have led to the destruction of the natural world. He suggests that recognizing the negative aspects of modern technology and science and developing suitable alternatives is necessary to address this problem.

Chapter 2

Nasr on the Roots of the Ecological Crisis and its Main Questions

In his work, Nasr posits that the current environmental crisis is a result of the scientific worldview that has come to dominate modern society. This worldview, based on the tenets of modern science and technology, has led to human actions that have caused the degradation of the natural world. Nasr's book is supported by the findings of many climate scientists, who warn of the dire consequences of continued environmental degradation. To address the issue, Nasr suggests that we must recognize the negative aspects of modern science and technology and develop appropriate alternatives that are more sustainable and responsible. In this chapter, we focus on Nasr's ecological message and its relevance to both Muslim and non-Muslim readers. We examine Nasr's San Rossore speech as the foundation of his primary ideas, supplemented by references to his previous works.

The chapter is divided into five sections, which evaluate the severity and actuality of the environmental problem, critique modernity and its contribution to the crisis, propose a change in perspective as a method for resolution, present the traditional worldview as an alternative, and suggest ways to bring about a transition or paradigm change. The chapter's objective is to highlight Nasr's key findings and provide historical justification where applicable. We begin by discussing Christian worldviews prior to the development of modern science, exploring how modern science supplanted these worldviews with its rationalist intellectual movement. Finally, we examine how the advanced scientific worldview has led to a contemporary human being who is driven by materialistic objectives and ambitions, leading to the environmental crisis. Nasr's critical evaluation of modern science and scientism is particularly relevant to Muslims, as it can shape their worldview and lead them to adopt a materialistic lifestyle that disregards the preservation of nature. By overlooking foundational principles in modern science, Nasr argues that the limitations of modern science become inherent, which can have significant consequences. Nasr's work is rooted in the teachings of the Qur'an, which emphasizes the importance of contemplating the creations of the heavens to serve as a gateway to inspiration. Traditional scholars, such as al-Iskandari and al-Ghazzali, underscore the need to recognize the greater realities of both the objective and subjective poles of nature as prerequisites for comprehending God's signals. Nasr highlights the importance of acknowledging both the intellectual capacity of the human subject, known as the "eye of the heart," and the transcendental aspects of the objective universe that can only be comprehended by a pure heart. Overall, Nasr's critique of contemporary science highlights the importance of recognizing the impact of our actions on the environment and developing alternative approaches that are more sustainable and responsible. His work emphasizes the need to transition towards a more traditional worldview that values the preservation of nature and the recognition of the greater realities beyond the physical world.

Root Causes of the Ecological Crisis of the Contemporary World

> The Earth is suffering from wounds inflicted by humanity, which has lost its harmony with the divine and is therefore in constant conflict with the environment. The destruction of nature is happening on an unprecedented global scale, caused by both those who have secularized the world through science and technology, and those who still live within a religious framework, though their methods of destruction differ. The environmental crisis is now a global issue affecting the entire planet. Surprisingly, despite the fact that the destruction of nature is primarily the result of a secular worldview, the majority of humans still live within a religious framework. Therefore, the role of religion in solving the crisis between humanity and nature is crucial. There is a need to develop a path that spans across religious boundaries without compromising the significance of religion itself. A comparative study of how different religions view the "Earth" is necessary......[226]

The opening line of Nasr's book "Religion and the Order of Nature" is analyzed in detail in this passage for its perfect encapsulation of Nasr's

[226] S.H. Nasr, "Religion and the Order of Nature" (New York Oxford: Oxford University Press, 1996), 3.

eco-philosophy. The line captures several key aspects of his argument, including a sense of urgency about the ecological crisis, the absence of harmony with nature, and the role of technology and science in causing the problem. The passage also highlights the importance of religion in solving the crisis, through traditional religious cosmology and metaphysics. Nasr argues that secular modernity is the root cause of the ecological dilemma and that a return to traditional religious understandings of nature is the solution. In sum, the opening line of Nasr's book serves as a powerful introduction to his eco-philosophy and the central arguments he makes throughout the book. Nasr views the "environmental crisis" as including the full spectrum of issues that are typically included under the "environment," such as species extinction, soil, water, and ocean pollution, climatic change or global warming, deforestation, and coral reef degradation. Nasr is aware of the truth and importance of these issues.[227]

> The environmental crisis is increasingly visible to those who acknowledge it. It's a daunting task for proponents of perpetual economic growth and linear human progress to deny the threat it poses to humanity, even with the support of a few scientists. Modern human activities on the environment have created and continue to fuel the crisis.[228]

> Our civilization is so out of balance with nature that it's arguably the most destructive weapon of mass destruction when taking a long-term perspective. The destruction of the environment also means our own self-destruction. In short, ecocide equals suicide.[229]

In his view, the symptoms of environmental decay are indicative of a single underlying issue: the breakdown of natural balance and harmony. The "environmental crisis" is essentially a crisis in humanity's relationship with nature. In one interview, he described a "crisis" as a state that is abnormal, hazardous, and imbalanced.[230] Nasr uses "environment" to refer

---

[227] The San Rossore Lecture was printed as "Man and Nature: Quest for a Renewed Understanding" in: Sophia, The Journal of Traditional Studies, vol. 10, number 2, winter 2004, p5-14.

[228] Ibid., 2004, p5-6.

[229] Sophia, vol. 10, no. 2 (2004): 14.

[230] Seyyed Hossein Nasr, interview by Muzaffar Iqbal, "The Islamic Perspective on the Environmental Crisis,"

to "natural" or "unspoiled nature" that excludes human-made or altered entities. This definition includes the "inner environment" of our bodies, which is also under threat due to food contamination. According to Nasr, it is evident that mankind's actions have taken a self-destructive turn. In his opinion, there is little doubt that we have embarked on a suicidal path.[231]

The destruction of nature poses a grave threat to our existence as a species, and if we continue to harm the environment, we will ultimately destroy ourselves. This is the essence of the viewpoint shared by Nasr and many contemporary scholars, including Al Gore, who gained international recognition with his film, Inconvenient Truth.[232] In order to avert a catastrophe, it is imperative that we take decisive action aimed at addressing the root causes of the crisis. While efforts such as better environmental engineering, the development of environmental ethics, and conservation practices are commendable and important, Nasr warns that they can only serve to slow down destructive tendencies. He emphasizes that protecting the earth and all its terrestrial creatures must be the top priority for governments, social groups, and individuals alike. There is no task more urgent or important than this. Nevertheless, even with all these efforts, humanity can only hope to delay the inevitable for a while longer. Therefore, it is crucial that we take the necessary steps to address the root causes of the crisis, including overconsumption, pollution, and environmental degradation. Only then can we hope to avoid the self-destruction that ecocide will inevitably bring?[233]

> Humanity is facing its greatest natural crisis due to the prevailing paradigm that has caused us to treat nature as a passive entity that can be exploited without consequences. This Promethean dream has turned into a nightmare, and we are like sleepwalkers heading towards the edge of a precipice. We must awaken from this dream of forgetfulness and seek the wisdom of humanity to understand our place in the world. Our well-being is linked to the health of the environment, and we must recognize this interconnectedness.[234]

---

231 Seyyed Hossein Nasr, "Traditional Man, Modern Man and the Environmental Crisis," interview by Agenda Viva, Sophia, vol. 12, no. 2 (2006): [page number not provided].

232 Sophia, Volume 10, Number 2, 2004, p.14

233 Ibid., p.14

234 Ibid., p.14

> A holistic approach that integrates science, ethics and spirituality is needed to shift away from the destructive paradigm towards a more harmonious relationship with nature. We have a responsibility to act sustainably and ethically for the sake of ourselves and the planet.[235]

Nasr advocates for a shift in our worldview, which he believes is flawed and responsible for the environmental crisis. He identifies the "modern paradigm" as the dominant worldview shaping Western civilization, and points to its evolution as the cause of a spiritual crisis for modern humanity. Nasr delves into the historical development of Western philosophy in his writings, highlighting how the modern worldview has replaced the medieval Christian worldview.[236]

The gradual transformation of Western culture was marked by significant events such as Renaissance Humanism, the Enlightenment, and the Industrial Revolution. The shift in Western perspectives towards nature, spirituality, and self differed greatly from pre-modern times due to certain events. While acknowledging the diversity of Western culture, Nasr identifies the 'scientistic' paradigm as a major influence on its development. This perspective has played a significant role in shaping the Western world over the past few centuries.

The current perspective on the world is one that is secular, materialistic and focused on the reduction of complex systems to simpler components, positivism, reason, and the centrality of human beings. The "modern" view, commonly known as the scientistic paradigm, views nature as a machine, and its influence has led to remarkable scientific and technological breakthroughs, including DNA and quantum physics discoveries, space exploration, military equipment, and bioengineering. These advancements have also given rise to societies with robust political and financial institutions, ethical guidelines, and high living standards that prioritize individual human rights. Nasr does not deny the benefits and advancements that have come from Western civilization and modern science and technology. However,

---

235 Ibid., p.8. The accent is mine.

236 See: "Philosophy and the Misdeed of Philosophy" in Nasr, S.H. "Religion and the Order of Nature," Oxford University Press, New York Oxford, 1996. Also: Chapter 1 "Knowledge and its Desacralization" in: Nasr, S.H., "Knowledge and the Sacred", Albany: State University of New York Press, 1989.

he does object to the idea that science alone can give us a complete and absolute understanding of reality. He believes that the perspective offered by modern science is limited and only provides a narrow view of the world.

"Scientism" is a perspective that is rooted in two essential flaws: the first one is its disregard for the inner nature of humans, and the second is its disregard for the spiritual truth of nature. Nasr has dedicated most of his life to discrediting this perspective, which he believes is fundamentally incorrect and contributes significantly to the "spiritual crisis." This crisis is characterized by a narrow vision of reality and a widespread disregard for its spiritual elements. Nasr identifies two problematic characteristics of the contemporary worldview: the first is the absolutization of the human state or anthropocentrism, and the second is the limited, reductionist, materialistic view of reality and nature.

First, humans have traditionally viewed themselves as the center of the universe and autonomous, with this perspective being referred to as "Promethean Man." In contrast to the traditional perspective, the modern view recognizes only limitations on human conduct imposed by other humans, rejecting any notion of revealed or divine law. Human-made laws and regulations are regarded as arising from particular social and historical conditions. They are also considered to be subject to change over time. According to the author, this shift in perspective has led to the neglect of humans' responsibilities and dependence on the rest of the natural world.

> From the Renaissance and into the seventeenth century, modern humans began to perceive themselves as exclusively earthly beings, absolved of responsibility towards both God and the natural world. The sanctity of nature was disregarded, and nature was viewed merely through quantitative and mechanical lenses.[237]
>
> Humans increasingly ignored not only God's rights but also the rights of other creations, while downplaying human responsibilities. Humans have adopted a mentality of being in control of their fate and the world, with little consideration for the inherent rights of God's creatures.[238]

[237] *Sophia*, Vol 12, No2, 2006, p.29.

[238] *Sophia*, Volume 10, Number 2, 2004, p.7

According to the current secular perspective, there is no fundamental or divine established reason for people to care for the other living things on the planet. Why should we care at all? Environmental ethicists have been debating this central topic for years since it lies at the core of all environmental ethics. To address this challenge, many strategies, including biocentric and ecocentric methods, were created. We shall see that Nasr will offer a theocentric or God-centered approach.

Second, the reduction of nature to only "quantity and mechanical relationships" has resulted in the loss of its intrinsic value and meaning. As a result, nature has been transformed into a mere "machine or quantity in motion," devoid of any qualities or spiritual reality.[239] This materialistic worldview has given rise to a science that emphasizes control and dominance over nature while disregarding its spiritual dimension. Unfortunately, this approach has led to imbalance and disharmony of the natural environment.[240] Modern science's perspective on the human being is similarly reductionist, viewing humans as nothing more than physical bodies made up of flesh and bones, electric currents, neural synapses, hormones, and DNA structure. This perspective denies the consciousness, psychic, and spiritual abilities of humans. Instead, it portrays humans as complex machines that have evolved from single cells to apes and, ultimately, to humans. The reductionist view of nature and humans in modern science has resulted in a loss of the intrinsic value and meaning of nature and has led to the destruction of the natural environment. Additionally, this perspective denies the spiritual and psychic dimensions of humans, reducing them to mere physical machines.

Now, what makes this "scientific" perspective flawed, and why does it believe that this is what is causing the environmental problem? The current economic structure, in which Nasr argues that the "rampant consumerism" is a manifestation of a "system of institutionalized greed" we are experiencing worldwide nowadays, is the primary source of the problem. According to the contemporary worldview, people are just "flesh and bones" and "economic animals," with all of their demands and

---

[239] *Sophia*, Volume 10, Number 2, 2004, p.11

[240] Ibid,. p.7

motivations being purely material in nature.[241] This perspective led to an economic framework that is primarily motivated by the "attraction to the emotion of greed".

> The present crisis is a result of the modern economic system that capitalizes on human passions, particularly the passion of greed, which is further exacerbated by the creation of false needs. This stands in contrast to the timeless religious doctrine of contentment, which emphasizes being satisfied with what one has. The modern perspective, on the other hand, feeds the flames of covetousness and greed, striving to attach the soul more firmly to the material world and turning a virtue – detachment from the world through asceticism – into a vice.[242]

> Happiness has become synonymous with material acquisition, while asceticism is now viewed as a "sin." This attitude is perilously close to the rampant consumerism that fuels never-ending desires that transform into "needs" for material goods, all of which must be satisfied by finite resources.[243]

So materialism is the direct reason, but the root cause is a mindset that completely ignores spiritual reality, which lies behind this most obvious cause. The current economic structure and the ensuing environmental damage are the external consequences of the modernist or "scientific" mindset. The problem of the inner spiritual crisis has been increasingly externalized as an environmental issue since the Industrialization era.[244] Nasr's work centers around his criticism of modernity, which permeates not only his environmental writing but also his other works. His critique challenges the underlying assumptions and values of the modern world and its scientistic philosophy and is not a wholesale rejection of modernity, but rather an engaged critique. This kind of criticism has a long history in the West, dating back to the Romantic movements of the 19th century, and

[241] Ibid., p.7

[242] Nasr, S.H. 'Religion and the Environmental Crisis.', in: Chittick, W.C. (eds) The Essential Seyyed Hossein Nasr, The perennial Philosophy series, Bloomington: World Wisdom, 2007. p.31-32.

[243] *Sophia*, Volume 10, Number 2, 2004, p7

[244] Interview *Agenda Viva*, in *Sophia*, Vol 12, No2, 2006, p29

later reemerging in the counterculture movement of the 1970s.[245]

Nasr draws inspiration from Western environmentalism and shares similar views with the environmental and anti-globalist movements of the 90s. He also shares a modernist critique with many Muslim thinkers. This critique is also a key aspect of the traditionalist perspective, which will be explored in future chapters.

Origin of the Environmental Crisis and Criticism of the Western World

Seyyed Hossein Nasr's book, "Theodore Roszak's The Making of a Counter Culture,"[246] published in 1969, brought him to the forefront of environmental philosophy. He is known for his contributions to the field of religious studies, philosophy, and environmentalism. In his book "The Knowledge and the Sacred: An Introduction to Philosophy,"[247] published in 1989, Nasr argues that the environmental crisis is rooted in the divorce of philosophy from metaphysics. Seyyed Hossein Nasr's "The Origins of the Environmental Crisis"[248] offers a comprehensive analysis of the underlying causes of the environmental crisis. Nasr argues that the roots of the contemporary ecological crisis can be traced back to the pre-modern Christian worldview, which viewed nature as a sacred manifestation of God's divine wisdom and beauty. Nasr contends that this worldview changed with the rise of Baconian progressivism, which reduced nature to a collection of material objects that could be exploited for human gain, leading to the divorce of philosophy from metaphysics.

Nasr contends that pre-modern Christian Europe had a fundamentally different relationship with the natural world than modern society. He argues that the Christian view of creation held that nature was inherently

[245] Hahn, L.E. Auxier, R.E., Stone, L.W. Jr, 'Preface' in: *The Philosophy of Seyyed Hossein Nasr*, 2001, pxvii

[246] Nasr, Seyyed Hossein. "Theodore Roszak's The Making of a Counter Culture." Review of Existential Psychology & Psychiatry, vol. 10, no. 1-3, 1972, pp. 197-207.

[247] Nasr, Seyyed Hossein. "The Knowledge and the Sacred: An Introduction to Philosophy." State University of New York Press, 1989.

[248] Nasr, Seyyed Hossein. "The Origins of the Environmental Crisis." Science & Spirit, vol. 7, no. 5, 1996, pp. 8-12.

valuable, as it was created by God, and that humans had a stewardship responsibility to care for it. In contrast, modern society sees nature as a resource to be exploited for human benefit and places little value on its intrinsic worth.

Nasr's book is supported by various scholars who argue that the mechanistic worldview of Baconian progressivism paved the way for industrialization, capitalism, and a culture of exploitation of the natural world. For instance, Bacon (1620) advocated for the domination and manipulation of nature for human ends,[249] while Descartes (1637) promoted a dualistic worldview that separated the physical and spiritual realms, reducing nature to a mere object to be studied and manipulated.[250] Furthermore, Habermas (1987) notes that the Enlightenment's emphasis on human rationality and progress resulted in the development of scientific technology that disregarded nature's intrinsic value.[251]The Baconian paradigm led to the industrial revolution and capitalism, where economic growth and profit became the primary goals at the expense of ecological health. This paradigm is also responsible for the exploitation, pollution, and degradation of the natural world.

Despite Nasr's book, some scholars have criticized it, arguing that economic development is necessary to address environmental concerns. Furthermore, Nasr argues that the separation of metaphysics from philosophy has led to the disappearance of the sacred aspect of nature. He suggests that traditional metaphysical systems viewed nature as a manifestation of the divine, and thus deserving of reverence and protection. However, modern philosophy, which prioritizes reason over faith, has largely abandoned this view. Despite criticisms of his views, Nasr's work has been vindicated by some economic historians. They suggest that societies that maintain a strong metaphysical orientation tend to have a more sustainable relationship with the natural world. For example, the medieval Islamic world maintained a strong sense of the intrinsic value of nature, and as a result, developed sustainable agricultural practices that allowed them to maintain a productive relationship with the natural world for centuries.

---

[249] Bacon, F. (1620). Novum Organum. London: J. Haviland.

[250] Descartes, R. (1637). "Discourse on Method." Paris: Henry Le Gras.

[251] Habermas, J. (1987). "The Philosophical Discourse of Modernity: Twelve Lectures." Cambridge: MIT

The State of Nature during Pre-Modern Christian Europe

To begin our examination of contemporary technology, we must first look at the Christian perception of nature, which was eventually replaced by modern science. Prior to the development of modern technology, many famous Christian philosophers depicted nature from a perspective that emphasized the presence of God in nature. [252] This perspective highlights the significant transition that occurred in Western civilization with the emergence of modern technology. Nasr points out that perennial metaphysical ideas, such as the unity and hierarchy of reality, as well as the purposefulness of the universe, can be found in classic Christian texts either directly or indirectly.

First, the idea of the oneness and omnipresence of God's essence is supported by the Bible, as well as notable figures such as Francis of Assisi, Hildegard of Bingen, and Meister Eckhart. This interpretation is further reflected in the concept of *vestigia dei*, which highlights the "signs of God" that can be found in all of nature's objects. Saint Paul's statement in Romans 1:20 affirms this notion, stating that God's invisible qualities have been evident in the world since the beginning of creation and can be understood through what has been made, demonstrating His eternal power and divine nature.[253] [254]

Second, in ancient Greece, St. Maximus (580-662) and other Church Fathers introduced the concept of logos, which suggests that Christ, also known as the Word of God, embodies the values of all beings. This concept also suggests that God created the universe through Christ.[255] Similarly, other Platonic and Neoplatonic Christian philosophers believed that all physical beings have archetypes or principles that exist in the Divine world. There-

---

[252] Philip Sherrard, *"The Rape of Man and Nature"* (Ipswich, UK: Golgonooza Press, 1987), 64. Also see Wolfgang Smith, *"Cosmos and Transcendence: Breaking Through the Barrier of Scientistic Belief"* (Illinois: Sherwood Sugden & Company, 1984), 45-60.

[253] Seyyed Hossein Nasr, *"Religion and the Order of Nature"* (New York and Oxford: Oxford University Press,1996), 56-59.

[254] Ibid., 136.

[255] Ibid., 58-59. The *"logoi"* of things is the "fundamental meaning" in accordance with which the thing is created by God. See Andrew Louth, *"Maximus the Confessor"* (New York: Routledge, 1996), 65.

fore, the logos concept and the idea of Divine archetypes or principles help to explain the interconnectedness of all things and emphasize the importance of Christ as a unifying force in the universe.[256] Third, since the 13th century, the fusion of Aristotelian cosmology and Christian philosophy by St. Thomas Aquinas (1224-74) has been a central aspect of scholasticism. This idea is illustrated in Dante's Divine Comedy, where the concept of a finite and geocentric cosmos with a hierarchy is used as a metaphor for the different levels of reality that the human soul must progress through in order to attain the ultimate reality of God – which is God's own reality.[257]

According to Nasr, the rise of rationalism in Western philosophy since the 12th century paved the way for modern technology. While admitting a variety of interconnected factors, he sees rationalism as the fundamental philosophical basis. Rationalism does not merely refer to the use of reason, but the exclusive use of reason without considering intellect and revelation. The Reason is considered the ultimate and exclusive authority for obtaining truth. In our previous discussion, we differentiated between reason and intellect. Thus, Nasr argues that rationalism's dominance in Western philosophy has played a crucial role in shaping modern technology.[258]

Undoubtedly, rationalism as a whole opposed the legacy of Meister Eckhart (1260–1327), St. Thomas Aquinas (1224–74), and St. Augustine (354–430) and their adherent.[259] Similarly, rationalism suppressed the holy or sapiential aspect of nature, which is only comprehended by Intelligence. According to Nasr, reducing Intellect to reason has not only made sacred knowledge inaccessible and meaningless, but it has also led to the destruction of natural theology. In the Christian context, natural theology reflected knowledge of sacred order and wisdom, which were lost due to this reduction.[260]

Nasr argues that by the late 17th century, the increasing dominance of rationalism in intellectual circles of the Christian West had a significant impact on how individuals perceived the universe and themselves. During

---

256 Nasr, *"Religion and the Order of Nature,"* 57 and 97-98.

257 Ibid., 99-100.

258 Nasr, *"Religion and the Order of Nature"*, 170.

259 See W. Norman Pittenger, "The Christian Philosophy of John Scotus Erigena," *The Journal of Religion,* vol. 24, No. 4 (Oct., 1944), 250.

260 Nasr, *Knowledge and the Sacred,* 4.

the post-medieval era, higher levels of reality were removed from both the subjective and objective domains. The human reason became the highest aspect of man, and the objective world was limited to what could be comprehended through normal human senses. This resulted in a reduction of spiritual and metaphysical aspects of reality, leading to a materialistic view of the world and human nature in the West.[261] According to Nasr's observation, this transition took place over the course of three main, gradual stages. The first stage, which took place after Toledo, a Moorish city, fell in 1085, was characterized by the influence of Islamic Aristotelian philosophy and related Islamic sciences on Christian thought.[262] Renaissance Humanism, a school of thought, served as the primary intellectual force behind the second stage. The separation of philosophy from metaphysics and revelation that took place during the first two stages, according to Nasr, provide the groundwork for the rise of contemporary technology in the 17th century. These changes in our view of the human self and nature are what Nasr claims.[263]

The third stage is in the 17th century, the Scientific Revolution occurred, marking the separation of science from metaphysics and revelation. The Revolution involved a total refusal of the timeless principles of reality's unity and hierarchy, as well as the ultimate meaning and purpose of the universe.[264] We will be discussing the philosophical foundations of modern science and their impact, with a focus on highlighting the key factors that led to the Scientific Revolution. The era was characterized by the dominance of rationalism and empiricism. Additionally, we will analyze the fundamental philosophical assumptions of modern technology and how they contradict conventional Christian worldviews. First, we will summarize Nasr's perspective on the contribution of Renaissance humanism to the Scientific Revolution, and then we will emphasize the importance of Francis Bacon (1561–1626), who, according to Nasr, epitomizes the aspect of modern science that seeks to dominate nature rather than comprehend its order.[265]

---

[261] Nasr, *Traditional Islam in the Modern World*, 100-101.

[262] Nasr, *Man and Nature*, 60-61.

[263] Nasr, *Religion and Order of Nature*, 177.

[264] Nasr, *Man and Nature*, 70.

[265] Seyyed Hossein Nasr, "Religion and Order of Nature" (New York: Oxford University Press, 2014), 135.

The Renaissance's humanist movement was multifaceted and often contradictory.[266] Unquestionably, one of the most significant movements was the rise of rationalist philosophy.[267] First, until well into the 17th century, Aristotelian philosophy was still taught in universities, and it was progressively taught outside of a Christian perspective.[268] Second, Nasr points out that many influential thinkers of the Renaissance, such as Giovani Pico (1463-1494), Lorenzo Valla (1406-1457), Marsilio Ficino (1433-1499), and Charles de Bouvelles (1475-1566), praised the human capacity for reason and freedom. These abilities enabled individuals to understand and experience the world, including God, without divine assistance.[269] Bouvelles even likened the wise person to Prometheus, a mythological figure who dared to steal the power of the gods. Another significant philosopher of the late Renaissance, Giordano Bruno (1548-1600), believed in the ideal of autonomy as part of the ideal of humanity. However, as autonomy became more important, it became increasingly separate from religion. [270] [271]

Along with developing rationality, another philosophical movement became prominent, notably in the 16th century: skepticism. Skepticism is not only applied to metaphysical facts but also to the basic possibility of knowing the truth. The interest in Sextus Empiricus' skepticism experienced a surge during the early 16th century (d. 3rd century), which was reinvigorated by the humanist movement.[272] Montaigne (1533-92), a prominent figure of the late Renaissance, was greatly influenced by the extensive study of Sextus Empiricus' skepticism. This happened around the same time when Nicolaus Copernicus (1473-1543) introduced his heliocentric model of the universe in 1543, which transformed the European perception of cosmology with the

[266] Lauro Martines, "The Protean Face of Renaissance Humanism," Modern Language Quarterly 51, no. 1 (1990): 105-121.

[267] Huston Smith, Cosmos and Transcendence: Breaking Through the Barrier of Scientistic Belief (Wheaton, IL: Quest Books, 2008), 44-47.

[268] David C. Lindberg, "The Beginnings of Western Science: The European Scientific Tradition in Philosophical, Religious, and Institutional Context", Prehistory to A.D. 1450 (Chicago: University of Chicago Press, 2007), 228-249.

[269] Nasr, "Religion and Order of Nature", 173-175.

[270] Ibid., 175-76; Ernest Cassirer, "*The Individual and the Cosmos in Renaissance Philosophy*", trans. Mario Domandi (Chicago: University of Chicago, 1963), 96.

[271] Ernest Cassirer, Ibid., 98.

[272] Seyeed Hossein Nasr, "*Religion and Order of Nature,*" 171.

help of mathematical reasoning.[273] Moreover, the rationalistic movement gained momentum by presenting a Promethean portrayal of the human being that surfaced during the Renaissance. This movement remained dominant over other schools of thought during the Scientific Revolution and the subsequent Enlightenment. Francis Bacon (1561-1626), a renowned figure in late Renaissance philosophy, held a strong rationalist stance in his understanding of the natural world, despite his outward religiosity. Bacon was the most notable proponent who provided a solution to the skepticism prevalent during the late Renaissance.

The Separation of Philosophy from Metaphysics and the Rise of Baconian Progressivism

Bacon had a crucial role in promoting the idea that understanding nature's material aspect and advancing mechanical arts to exploit its power was a certain way to progress human civilization. This philosophy of human advancement only reached its full potential after the Technological Revolution of the 17th century. In the early 1600s, Bacon's proposed epistemology for creating practical arts had the potential to revolutionize the prevalent approach of exploring the specifics of the observable world through the filter of metaphysical concepts:

> There are only two ways of seeking and revealing the truth. The first approach involves departing from the senses and particulars and moving towards the most general axioms. By assuming the truth of these principles to be settled and unchanging, one can arrive at judgments and the discovery of intermediate axioms. Currently, this approach is in vogue. The second method involves deriving axioms from the senses and particulars, gradually and continuously ascending until the most general axioms are reached. This is the correct approach, but it has not been attempted yet.[274]

Our goal is more effectively served by breaking nature down into its

---

[273] Nauret, *"Humanism and the Culture of Renaissance"*, 218.

[274] Francis Bacon, "The New Organon," Book One, Aphorism no. xix in *"The New Organon and Related Writings"*, ed. Fulton H. Anderson (New Jersey: Prentice Hall, 1960),43.

component pieces; Instead of focusing on forms, we should focus on the matter, its formations, changes in structure, basic actions, and principles of action. Forms are products of human cognition unless you prefer to refer to them as principles of behavior.[275] Bacon is alluding to the persistent Aristotelian concept of hylomorphism, which posits that every physical entity consists of both "form" and "matter." In this view, "form" represents the entity's essence and serves as the true subject of understanding, while "matter" is the inexplicable potentiality.[276] Although Bacon aimed to give natural philosophy a more materialistic orientation than his predecessors, he followed in the footsteps of the famous late Renaissance natural philosophers by rejecting Aristotelian forms and relying solely on practical knowledge.[277] Therefore, Nasr's contention that Renaissance Humanism separated philosophy from metaphysics and revelation is well corroborated by Bacon.[278]

The central idea of Bacon's approach was that studying accidents alone, rather than general laws or abstract Aristotelian forms of things, could lead to the cumulative knowledge of nature. Theodore Roszak illustrates the differences between information acquired through Baconian and conventional approaches.[279] According to Roszak, no other field of knowledge, including philosophy, theology, or literature, could be accumulated in a similar way since their fundamentals are immaterial and require the use of emotions, thoughts, beliefs, and inferences, in addition to reason. According to Roszak, while it's possible to gain insights from another person's interpretation of Shakespeare, Plato, or the New Testament, one cannot simply build upon their understanding. Instead, it's crucial for individuals to delve deep into these texts and comprehend them based on their own experiences and insights. Roszak believes that the traditional arts and philosophy are founded on perennial wisdom and that this wisdom can only be accessed

---

[275] Ibid., 53.

[276] Peter Dear, "*Revolutionizing the Sciences: European Knowledge and Its Ambitions", 1500-1700,* 2nd edition (Princeton: Princeton University Press, 2009), 64.

[277] See Fulton H. Anderson, Editor's Introduction to "*The New Organon and Related Writings,*" pp. xii-xiii. Also see Cassirer, *The Individual and the Cosmos in Renaissance Philosophy,* 145-46.

[278] Bacon, "The New Organon," Book one Aphorism no. LXV, p. 62.

[279] Theodore Roszak, "*Where the Wasteland Ends: Politics and Transcendence in Postindustrial Society"* (Berkeley, CA: Celestial Arts, 1989), 142-59.

through personal engagement with the texts.[280] Therefore, an individual's interpretation of Shakespeare's Hamlet can reveal as much about their own intellect as it does about the play itself. According to Bacon, if only chance occurrences hold significance, then someone else could easily take over where one left off and finish the task at hand. The same cannot be true, however, for someone who records changes in the accidents connected with natural phenomena.[281] Such intelligence was based on numerical measures, which could be understood without the use of intuition or contemplation and simply required the normal logical faculty to work.

According to Bacon, the universal rules for the purely physical origins of natural events may be inferred from the gradual development of physical facts on specifics. Such understanding would provide people with the ability to harness the force of nature, advancing human civilization. Bacon asserts that the goal of his scientific pursuit is not to create mere arguments but rather to develop practical applications that can manipulate nature to achieve desired outcomes. This objective is focused on tangible results and disregards the abstract concepts and theories of philosophers.[282] [283]

Bacon believed that his empirical and cumulative approach to knowledge acquisition, as outlined in his work "The New Organon," should be applied to all fields of study except those related to belief in God.[284] He maintained that human progress should be based on practical principles while leaving matters of faith untouched. This philosophy is known as "Baconian progressivism," and it suggests that progress can be achieved through the grounding of all intelligence on pragmatic basics without affecting one's faith.[285] [286]

Nasr argues that, prior to the modern era, the primary objective of Western Christian civilization (and all civilizations) was the attainment of moral

---

280 Ibid., 154.

281 Bacon quoted in Roszak, *"Where the Wasteland Ends"*, 150.

282 Francis Bacon, "The Great Instauration," in *"The New Organon and Related Writings"*, ed. Fulton H. Anderson (New Jersey: Prentice Hall, 1960), 19.

283 Bacon, "The New Organon," Book One, Aphorism no. LXXXIV, p.80.

284 Bacon quoted by Sherrard, *Rape of Man and Nature*, 68.

285 Ibid.

286 Bacon, "Great Instauration," 14-15.

virtue, rather than material progress as advocated by Bacon. Historian Torben Nielsen suggests that during the Renaissance, the values of "happiness" and "moral virtue," which were once considered supreme blessings, were reduced to secondary considerations in the pursuit of individual freedom.[287] [288]

Baconian progressivism's emergence is understandable in light of the historical setting.[289] The general people were living in abject poverty, and the obvious benefit of the growing technological skills in metallurgy, mining, mechanical, engineering, clocks, etc., was seen as a way to improve the general people's miserable material circumstances.[290] Nevertheless, despite his possibly good intentions—he believed he was on a divine assignment to materially elevate humanity—he showed little care for the wellbeing of the rest of nature. In any event, Bacon aimed to refute any claims made in the Christian faith about the hierarchical pattern of reality and the idea that nature is a vision of God's signs by emphasizing scientific certainty. He did this by directing human endeavor toward material advancement. Charles Taylor, a well-respected philosopher, stated that the Baconian perspective relinquished the endeavor to interpret the universe as a place of symbols and instead embraced an instrumental approach to achieve effective results.[291] Baconian epistemology holds that objects have neither an essential purpose nor an ultimate cause, in contrast to the dominant scholastic worldview of the time.[292] Overall, if Bacon considered his approach would accomplish God's plan for people, he constrained it to secure their materialistic well-being.[293]

Nasr suggests that the notion of material progress is the primary goal of life stems from two main influences: firstly, the reduction of humanity to a purely material entity, and secondly, a materialistic utopianism that

---

[287] Nasr, "*Need for a Sacred Science*", 150-151.

[288] Torben Nielsen, "The State, the Market and the Individual," *Acta Sociologica,* 1986 (29), 4:283-302), 284.

[289] Lauro Martines, "The Protean Face of Renaissance Humanism,"105-106.

[290] William R. Shea "The Scientific Rev Really Occurred" *European Review,* Vol. 15, No. 4, (459-471), 2007, p. 463.

[291] Charles Taylor, "*A Secular Age*" (Cambridge: Harvard University Press, 2007), 98.

[292] Ibid., 98.

[293] Ibid., 222.

can be traced back to Bacon.[294] Nasr uses the phrase "reduction of man" to describe how man has turned away from his superior nature, intelligence, and the more complete or elevated world that he might view via it. Therefore, to the level that the Promethean human rejected the conventional viewpoint that could be best comprehended by his intelligence, he also rejected the pursuit of spiritual perfection in favor of advancement in worldly pleasure.[295] By the conclusion of the 17th century's Scientific Movement, the worldly perspective of the Promethean human and the Baconian progressive ideology would have a solid rationalistic philosophical base and considerable reputation. In order for the Baconian argument to gain widespread acceptance, a new cosmology that rejected the concept of a hierarchical reality was necessary to fully secularize the universe. It would seem that this is exactly what took place throughout the Technological Revolution and the subsequent Age of Enlightenment.

According to Nasr and several other scholars, the Scientific Revolution thoroughly secularized nature.[296] In addition, the progress of secularization achieved various aspects of the Baconian vision by (1) altering the definition of "matter" from an incomprehensible and absolute potentiality to a measurable quantity, (2) dismissing structures or substances as suitable objects of inquiry, (3) establishing the philosophical basis (through Descartes' partition of the human self and any object of knowledge) for recognizing the possibility of a genuine empirical intellect that is free of human characteristics, and lastly, (4) introducing scientific laws and postulating them as the primary laws of nature.

Galileo (1564-1642) embraced the philosophical position that an object's quantitative features constitute its "primary qualities," while all other attributes are considered "secondary qualities." According to Nasr, this signified a deviation from the conventional notion of "matter" as pure potentiality, leading to the abandonment of the Aristotelian principle of hylomorphism, which was formerly supported by the Catholic Church. Moreover, this transformation substantiated Bacon's denial of the existence

294 Ibid., 153.

295 Nasr, "*Need for a Sacred Science*," 150-51.

296 Roszak, "*Wasteland*," 172-74, 178-80; Sherrard, "*Rape of Man and Nature*", 68-71; Smith, *Cosmos and Transcendence*, 13-30.

of substances and forms of objects within the scientific community.[297]

On the other hand, René Descartes (1596-1650) was impacted by Bacon via his mentor, Isaac Beeckman (1588-1637), a Dutch mechanical engineer. [298] Descartes held the conviction that the fundamental quality of the body was not characterized by traits such as solidity, weight, or color, but rather by its identity as a substance that extended in length, breadth, and depth. He termed this concept "*res extensa,*" meaning "extended thing." [299] He asserted that the attributes which Galileo had termed as "secondary qualities" were, in fact, created by the "*res cogitans,*" which he recognized as the human self. It is worth highlighting that the faculties of "res *extensa*" and "*res cogitans*" are entirely separate from each other. [300] Descartes also viewed matter as simply a collection of particles occupying space in three dimensions, and every observable phenomenon as the result of matter and motion. In a universe where even living beings were seen as functioning like machines, Descartes believed that motion was merely the result of collisions or frictions between different "extensions."

Descartes went one step further than Galileo by dividing the *res cogitans* and *res extensa*. Four years after Galileo's trial and conviction, in 1637, his *Discourse on Method* contributed to the development of the first rationalistic school of thought. According to this school of thought, human judgments and perceptions are ultimately governed by the capacity of reason, which is independent of any higher principles. Nasr argues that Descartes positioned the individual ego's thinking at the core of reality and as the yardstick for all knowledge, resulting in a philosophy of pure rationalism. This perspective placed the subject who knows within the realm of reason and separated them from both wisdom and revelation. As a result, these sources were no longer considered viable means for obtaining objective knowledge.[301]

---

[297] Based on notes from the course *Man and Nature* offered by Nasr at the George Washington University. 24 February 2004.

[298] *Philosophical Studies, C.1611-C.1619,* Vol. 6, edited by Graham Rees and Lisa Jardine (New York: Oxford University Press, 1996), p. lxxvii.

[299] Nasr, *Religion and the Order of Nature,* 139.

[300] Smith, *Cosmos and Transcendence,* 29.

[301] Nasr, *Knowledge and the Sacred,* 41-42.

In addition to restricting the universe to a material substance, similar to Galileo's approach, Descartes also confined the human self to the limits of common reason.[302] This ultimately lent philosophical validity to the Baconian pursuit of knowledge and the rationality of Promethean humans. Furthermore, Nasr observes that modern technology was significantly altering the traditional understanding of the "order in nature" across all societies. This shift was characterized by the replacement of the belief in divine laws governing the universe, both morally and otherwise, with a focus on mechanical laws of the material realm that could be uncovered through scientific experimentation.[303] According to Lawrence Schmidt and Scott Marratto, scrutinizing the physical world can lead to the discovery of its principles, but it falls short in illuminating the significance and intention behind human liberty, which necessitates philosophical reflection.[304] Both the classical and Christian traditions hold that the absolute natural order is the fundamental metaphysical reality, and behaving in accordance with it is integral to appropriate human conduct.[305] However, contrary to this conventional idea, Descartes' Discourse on Method identifies mechanical principles as natural principles from God. As a result, mechanical principles, as described through mathematics, came to be viewed as the laws of nature.[306] This perspective overlooks the fact that mathematical principles unless presented symbolically, do not have any relevance to the moral or other guiding principles of the universe.[307]

The transition from the conventional understanding of the universe to the new science was not an easy one for most people. The rigidly rationalistic and mathematical nature of the new science made it difficult for them to comprehend it for decades.[308] This was especially true for those who found meaning in their lives through their conventional beliefs. Historian of

---

[302] Smith, *Cosmos and Transcendence,* 28.

[303] Nasr, "Religion *and the Order of Nature"*, 132-33.

[304] Schmidt and Marratto, *"End of Ethics in a Technological Society"*, 15. Smith, *Cosmos and Transcendence,* 54.

[305] Ibid.15; see Lindberg, *"Beginnings of Western Science"*, 213.

[306] Nasr, *"Religion and the Order of Nature,"* 133.

[307] John L. Heilbron, "Coming to Terms with the Scientific Tradition," *European Review,* Vol. 15, No. 4, 2007, p 478.

[308] Margaret Jacob, *"Scientific Culture and The Making of the Industrial West"* (New York and Oxford: Oxford University Press, 1997), 18-27.

science, Sherwood Taylor, noted that prior to science becoming the sole legitimate way of comprehending nature, the vision of God in nature was the norm. It was not seen as an exceptional experience.[309] Galileo and Descartes were among those who attempted to promote their science by highlighting its practical benefits rather than appealing to people's higher natures or intelligence, as Nasr suggested. They focused on how their ideas could be used to improve people's lives in tangible, physical, or economic ways, much like Bacon did.[310]

Galileo and Descartes compromised mathematical findings as certainties in their search for certainty in an era of skepticism. By divorcing science from metaphysics or religion, Galileo and Descartes paved the path for the quantification and mathematization of nature with the aim of making it manageable. Now, for the first time ever, the use of nature for human material benefit was philosophically justified.[311]

Nasr claims that the Newtonian science that followed did not significantly change the fundamental philosophical ideas of the new science established by Galileo and Descartes.[312] Newton (1642–1727) did not in fact raise any objections to the concepts of quantification or Cartesian bifurcation. He believed in a cosmos that could be described mathematically and placed a strong emphasis on empirically established truths. Newtonian physics only had his own interpretation of spirituality, even if he recognized God as the origin of the rules of his physics. His mathematical formulations' definition of a mechanical worldview and the Cartesian dualism they included, which allowed for the imprinting of mathematical connections on the cosmos, is his legacy. John Henry, a renowned scientific historian, explains it pretty eloquently:

> By merging Bacon's emphasis on experimentation with Descartes' focus on rationalism, Newton was able to create a new understanding of reality that separated nature from the Platonic concept of an intelligible world, as espoused by St. Augustine, St. Maximus,

---

[309] Sherwood Taylor, "*The Fourfold Vision*" (London, 1945), 91; quoted in Nasr, "*Man and Nature,*"41.

[310] Jacob, "*Scientific Culture*", 18, 43.

[311] See, "*Science and Technology in World History,*" 245-46.

[312] Nasr, "*Religion and the Order of Nature,*" 140.

> and Johannes Scotus Erigena (810-877). This view also dissociated nature from the moral principles that governed human life, as envisioned by St. Thomas, and from any shared spiritual reality between humans and nature, as described in the doctrines of the unity and omnipresence of the Spirit and Christ as the Logos.[313] According to Wolfgang Smith, the modern scientific perspective represents a complete deviation from the Christian worldview. The Scientific Revolution of the 17th century is deemed revolutionary not only due to the unparalleled progress of Renaissance science but also because it fundamentally broke away from the Christian metaphysical conception of nature. This replaced the Christian understanding of nature with a mechanical interpretation that severed nature from the soul and eradicated any indication of a divine presence or consciousness.[314]

Newtonian science and its philosophy extended across the West throughout the Enlightenment and the subsequent Industrialization, and later in the places the West had conquered. The application of rationalistic and empiricist modern scientific methodologies to philosophy and the study of human civilization, making these fields as rational and "progressive" as modern science, can be traced back to the writings of John Locke (1632-1704). [315] [316] This trend gained momentum in the early 19th century, thanks to scientism, which advocates that current science or scientific reasoning is the most reliable method, if not the only method, for acquiring genuine knowledge.[317] Scientism was first restricted to a relatively small part of the intellectual elite up until the late 17th century, but it eventually swept over all of Europe.[318] Scientism had the implication that the scientific perspective

---

[313] Ibid 56, 59.

[314] Smith, "*Cosmos and Transcendence*," 47.

[315] Locke's atomistic conception of ideas. John Locke, "Of Innate Notions," in "*The Age of Enlightenment: The 18th Century Philosophers*," ed. Sir Isaiah Berlin (New York and Toronto: The New American Library, 1956), 36-45.

[316] "*The Quantifying Spirit in the 18th Century*," eds. Tore Frangsmyr, J.L. Heilborn, Robin E. Rider (Berkely: University of California, 1990).

[317] See Huston Smith, "Scientism: The Bedrock of the Modern Worldview," "in *Science and the Myth of Progress*", ed. Mehrdad M. Zarandi (Bloomington, IN: World Wisdom, 2003), 233.

[318] Richard G. Olson, "*Science and Scientism in Nineteenth-Century Europe*" (Urbana: University of IL, 2008).

of the time was legitimate. As a result, the Enlightenment was established on the Newtonian mechanical viewpoint. The ideology of scientism results in two key outcomes: (1) It eliminates the belief in deeper layers of reality beyond the tangible and material components of nature, and (2) It secularizes human consciousness. Jane Bennett provides a brief explanation of the impact of this ideology on nature, which has been evident since the Enlightenment. During the 18th century Enlightenment, there was a concerted effort to remove mysticism from the world in accordance with faith. Nature was viewed as a divine text written by God, filled with symbols, meaning, and order. In response to the belief in an enchanted universe, the Enlightenment attempted to distance God from a more immediate social position.[319]

Christianity faced a significant threat from Scientism, particularly following the Enlightenment, as noted by the distinguished philosopher Louis Dupre, who argues that the eighteenth century was the first non-Christian era. During this time, most prominent intellectuals and artists, even if they did not oppose Christianity, no longer drew inspiration from it, and secularism began to dominate. The nineteenth century was even more unfavorable for religion, characterized by an intense antitheistic campaign to eradicate all Christian influences from culture.[320] [321] The conventional view of the hierarchy of reality was strongly challenged by scientism during this time, exemplified by Marxism's claim to scientific truth, Darwin's theory of evolution, and Freud's attempt to construct a scientific psychology that disregards the spiritual realm.[322] [323] Even today, many academics unequivocally contend that modern science and technology have a secularizing effect.[324]

The philosophy of Baconian progressivism, developed by Sir Francis Bacon, championed the idea of progress through material evolution. It was a radical

[319] Jane Bennett, *"Unthinking Faith and Enlightenment: Nature and the State in a Post-Hegelian Era"* (New York: New York University Press, 1987), 7.

[320] Louis Dupre, "Spiritual Life and the Survival of Christianity," *"Cross Currents;"* Fall 1998, Vol. 48 issue 3, 381.

[321] Ibid.

[322] Olson, *"Science and Scientism in Nineteenth-Century"*, 163.

[323] Huston Smith, "Scientism: The Bedrock of Modern Worldview," 234.

[324] See George Kateb, "Locke and the Political Origins of Secularism," in *Social Research*, Vol. 76: No 4: Winter 2009, 1004-1005. (accent ours)

departure from traditional ways of thinking, which had focused on religious and metaphysical explanations of the world. Bacon believed that the pursuit of knowledge through empirical observation and experimentation would lead to advancements in science, technology, and society.[325] However, with the Enlightenment, there was a shift in focus from Baconian progressivism to scientism-centered progressivism. Scientism emphasized the role of science as the primary source of knowledge and truth, and it placed an emphasis on the quantification of nature. This new philosophy was founded on the belief that nature could be reduced to a simple quantity, which made development in the material sense inevitable. One of the key developments of this new philosophy was the rapid development of the experimental sciences. These sciences focused on the sensible world, the world of matter, and the world that had the most direct practical applications. This led to a surge of innovation in modern civilization, which had not been seen in any other society. [326] The reason for this scientific advancement lies in the fact that science limits reality to the material level. By dismissing the idea of higher realities of beings, scientists were able to accumulate knowledge about nature and develop technologies that had a profound impact on society. This philosophy is identical to Bacon's argument, which stated that the pursuit of knowledge through empirical observation and experimentation would lead to advancements in science and society.

As previously stated, the emergence of scientism had a secularizing effect on human consciousness. According to Nasr, this led people to focus exclusively on earthly pursuits, as the concept of the afterlife receded into the distance.[327] Since Bacon's time, the ideal of moral perfection has further degraded. In the period of enlightenment that followed the Scientific Revolution, human attention shifted from "being good" to "being happy" in the concrete, materialistic sense, as highlighted by the prominent historian Roy Porter.[328] Jacob Needleman, a renowned philosopher, viewed the same scenario as the result of modern technology depriving

[325] Nasr, "*Need for a Sacred Science,*" 150-51.

[326] René Guénon, "*The Crisis of the Modern World,*" trans. Marco Pallis, Arthur Osborne and Richard C. Nicholson (Hillsdale, NY: Sophia Perennis, 2001), 47.

[327] Nasr, "*Need for a Sacred Science,*" 151.

[328] Roy Porter, "The Enlightenment in England", in Roy Porter and Mikulas Teich, eds., "*The Enlightenment in National Context*" (Cambridge: Cambridge University Press, 1981), 14.

the universe of divine consciousness, which prevented it from inspiring in people the development of consciousness required for perception of the good and holy: Originally, science started as an exploration to discover a fresh perspective on how to comprehend the significance of the natural world. However, it eventually devolved into a practice of meticulously tallying punctuation marks. Over time, the urge to control nature became the primary focus, leading to the birth of pragmatism. Consequently, the objective of acquiring knowledge shifted from expanding consciousness to fulfilling personal desires.[329]

The ideology of scientism promotes a secularized view of both nature and humanity. As Nasr points out, this secular society is driven to exert increasing control and exploitation over the environment through the use of science and technology. Unfortunately, this approach disregards the significance and purpose of nature beyond its material existence, resulting in its continual degradation.[330] Nasr argues that as nature and human awareness become more secularized, people will rely more heavily on methods to explore and satisfy their desires within the only available domain—the secularized natural world. Consequently, modern technology, with its emphasis on rationalism and efficiency, has become the most reliable means of human progress. This perspective, referred to as "scientific progressivism," holds that modern technology and associated scientific reasoning are the primary—if not the only—means of advancing humanity. In summary, Nasr contends that scientism promotes scientific progressivism, leading to environmental crises and exploitation of the natural world. While it has taken time for the detrimental effects of scientism and scientific progressivism to become apparent, their impact is now undeniable. According to Schmidt and Marratto, the Enlightenment worldview, with its outwardly focused mechanistic philosophy void of spiritual meaning, attempted to eradicate or minimize outside sources of human misery – labour, poverty, illness, and conflict – and "Understood ethics within the context of this endeavor."[331] The Baconian approach to improving human conditions has overlooked the internal or spiritual aspi-

[329] Jacob Needleman, "*A Sense of the Cosmos: Scientific Knowledge and Spiritual Truth*" (New York: Monkfish Book, 2003), 36-37 and 17-21.

[330] Nasr, "*Religion and the Order of Nature,*" 179.

[331] Schmidt and Marrato, "*End of Ethics in a Technological Society*", 15.

rations of individuals, as well as the role of the spiritual viewpoint of nature in fulfilling those desires. The focus on manipulating the natural world for human gain has overshadowed the deeper, more intrinsic values that drive our existence. Consequently, the 20th century witnessed two devastating World Wars, followed by the emergence of a growing environmental crisis in the 1960s. These catastrophic events have forced a reconsideration of the scientism ideology and scientific progressivism.

Since the early 20th century, traditional Newtonian science, which is built on Cartesian philosophy, has faced significant challenges from various scientific developments such as the theory of relativity and quantum mechanics. Despite these challenges, it continues to be a dominant force in most scientific fields and remains deeply ingrained in the public consciousness of contemporary civilization. In recent years, there has been a growing understanding that there exists an actual reality beyond human perception, and technological advancements have allowed us to glimpse into this reality. According to John Carvalho, this viewpoint considers the capacity of the human mind and senses to comprehend this reality completely.[332] Nonetheless, Wolfgang Smith proposes that the extensive collection of physical theories resulting from the technological progress in the 20th century still heavily depends on traditional Newtonian principles. In essence, modern science still rests on the venerable Cartesian doctrine. Moreover, despite the significant developments in science and technology, scientific intelligence continues to remain "objective" in its perception of being separate from God or the soul. This perception stems from the Cartesian view that sees the human mind as a separate entity from the external world. In other words, scientific knowledge is perceived as being independent of any religious or spiritual beliefs.[333] However, according to Nasr, modern science is limited to dealing with external phenomena, no matter how much it changes. Despite the incredible progress made in fields such as particle physics and cosmology, science can only address the physical aspects of reality. In contrast, traditional religious and spiritual beliefs can provide a deeper understanding of the inner nature of reality.[334]

---

[332] John J. Carvalho IV, "Overview of the Structure of a Scientific Worldview," *Zygon,* vol. 41, no. 1 (March 2006), p. 121.

[333] Smith, "*Cosmos and Transcendence,*" 40.

[334] Nasr, "*Religion and the Order of Nature,*" 152.

Nasr's claim that the ecological problem arose due to the shift from a religious or metaphysical standpoint to a modern scientific outlook is supported by several people, such as Theodore Roszak, Philip Sherrard, Huston Smith, Wolfgang Smith, many Islamic environmentalists referred to in the Introduction, and several environment and religion scholars.[335] [336] [337] [338] Renowned scholars in the field of environment and religion, Mary Evelyn Tucker and John Grim argue that the ecological catastrophe was brought about by the perspective that has seized the imagination of present-day industrialized societies.[339] Nonetheless, in recent times, some of the most distinguished economic historians have recognized that the contemporary technological perspective itself played a pivotal role in shaping the present-day economic and industrial society, which constantly strives for advancement.

Nasr's Ideas Validated by Economic Historians

Nasr's assumptions have essentially been confirmed by historical evaluations by two renowned economic historians, Joel Mokyr of Northwestern University and Margaret Jacob from the University of California at Los Angeles (UCLA), which relate the transition in Western conceptions of environment in the 17th century to the Industrialization and the subsequent expansion of technology and the economy after it. We shall therefore go into extensive detail on the relevant works of these two academics.

In his recently published book "The Enlightened Economy", Mokyr asserted that economic change in all eras is influenced more significantly by people's beliefs than most economists commonly acknowledge.[340] According to

[335] Roszak, "*Where the Wasteland Ends,*" 231-237.

[336] Philip Sherrard, "The Desanctification of Nature", in "*Seeing God Everywhere: Essays on Nature and the Sacred,*" ed. Barry McDonald (Bloomington, IN: World Wisdom, 2003), 109-130.

[337] Smith quoted on dust jacket of Nasr's "*Religion and the Order of Nature*" (New York and Oxford: Oxford University Press, 1996).

[338] See Wolfgang Smith, "*Cosmos and Transcendence,*" 143-44.

[339] Mary E. Tucker and John Grim, "Series Forward," in "*Christianity and Ecology*", eds. Dieter Hessel and Rosemary Ruether (Cambridge: Harvard University Press, 2000), xxv.

[340] Joel Mokyr, "*The Enlightened Economy: An Economic History of Britain 1700-1850*"

Jacob and Mokyr, Industrialization was not merely the natural result of people developing the necessary technology to employ resources and other economic conditions to enhance their level of living materially. The velocity and extent of technical advancements served as the stimulus for rapid economic development. Mokyr implies that previous explanations for the Industrial Revolution, which predominantly focused on economic factors, have failed to provide a sufficient explanation due to the endogenous growth problem. None of these approaches can fully account for the phenomenon without acknowledging the role of technological advancements.[341] Scholars have usually argued that, with the exception of a few instances, 17th-century science had little direct impact on the advancement of technology in the 18th century.[342] However, what led to the many ongoing scientific advancements during the 18th century and afterwards?

Mokyr argues that the rise of a spirit of progression and advancement during the Age of Enlightenment was the driving force behind scientific developments. This culture, according to Mokyr, was defined by what he terms "Industrial Enlightenment," which he defines as "a belief in the possibility and desirability of economic progress and growth through knowledge."[343] This idea, according to Mokyr, may be traced back to the Baconian theory of social progress, in which "knowledge" is defined as information that is "useful" in harnessing the resources and power of the environment.

According to Mokyr, the price of gaining essential knowledge, such as scientific and industrial knowledge, was reduced, which helped to sustain the Industrial Enlightenment.[344] Even though Mokyr does not explicitly attribute the Industrial Revolution to the modern scientific worldview, it is implied in his explanation of the Baconian conception of useful knowledge. According to Mokyr, the remarkable economic advancements witnessed in the 18th century were due to the evolution of the definition of useful knowledge in the intellectual sphere. He suggests that this knowledge was

---

(New Haven: Yale University Press, 2009), 1.

341 Joel Mokyr, "The Intellectual Origins of Modern Economic Growth," *The Journal of Economic History*, Vol. 65, Number 2, June 2005, pp. 287-88.

342 See James E. MeClellan III and Harold Dorn, "*Science and Technology in World History*" (Baltimore: The Johns Hopkins University Press, 2006), 269.

343 Mokyr, "The Intellectual Origins," 291.

344 Ibid., 297-322.

not confined to scientific research, but also involved the gathering of information and techniques related to the physical realm.[345]

In Bacon's view, useful knowledge will continually expand and be made up of empirical evidence. Philosophically, this implied a change from a comprehensive to a fragmented perspective of reality; in other words, from the conventional viewpoint wherein higher realities were ignored and human actions were focused on learning about the multiplicities in the material realm alone to one in which higher realities were ignored and showed how integrally related the reality of the material world was to them. Mokyr acknowledges that the division of knowledge, characterized by specialization, professionalization, and expertization, is essential for the advancement of useful knowledge. This concept was already present in Bacon's fragmented view of reality, and his rejection of conventional epistemology is linked to the intellectual development of useful knowledge, which Mokyr believes is necessary to understand its evolution. [346]

During the Enlightenment, the Baconian perspective on progress in knowledge and wealth gained wide acceptance as a result of the scientific advancements made during the Scientific Revolution. The scientific revolution was characterized by a relentless focus on empirical evidence, rational thought, and experimentation. These intellectual pursuits led to the development of a science that was solely based on the physical realm. In contrast to Mokyr, Jacob puts forward a more direct claim that the contemporary scientific paradigm itself played a crucial role in establishing useful knowledge during the eighteenth century, and served as a catalyst for further progress in science and technology. According to Jacob, Bacon's religious aspiration to create a utopia for the less fortunate was overshadowed by the secular atmosphere of the eighteenth century, which was heavily influenced by Newton's mechanical philosophy.[347] Jacob argues that the utopian ideal simply served as a justification for technological advancement rather than a true motivation for social change. In fact, it was a new scientific understanding of nature that played a vital role in the development of the mechanized industry. This understanding led to the creation of new machines

---

[345] Ibid., 290.

[346] Ibid., 287.

[347] Jacob, *"Scientific Culture,"*33.

and tools, which in turn helped to make mass production possible, leading to greater prosperity and innovation.[348]

Jacob and Mokyr share the view that Newton's scientific work was simplified and widely disseminated to increase accessibility to new scientific knowledge. However, Jacob argues that Newton's Principia did not only pertain to planetary motion but also included information on the mechanics of local motion. This information was used by scientists to develop and merchants to consume curricula and books that had practical applications for technological innovation. As a result, the study of Newtonian physics from these sources produced individuals like James Watt, who was the inventor of the first steam engine. These individuals were part of teams consisting of mechanics and engineers who applied the knowledge gained from studying Newton's work to drive technological progress. Therefore, it's clear that industrialization was directly facilitated by modern science in the form of Newtonian mechanics.[349]

Newton's mechanistic worldview, according to Jacob, dramatically impacted people's interests and behaviors. By utilizing their understanding of Newtonian physics, engineers and the companies that employ them have been able to create technological advancements that allow them to objectify the physical world, perceive its operations in a mechanical manner, and incorporate their shared interests and values into their collaborations. This approach stems from a perspective that views the cosmos as a symmetrical mechanical system that functions based on the physical principles uncovered by Newton, rather than as a miraculous creation of God.[350] The human perspective toward the environment could only go from one of wonder due to our dependency on it for survival to one that is more exploitative if nature does not contain the signs of God or, as per Cartesian philosophy, have any relevance to us humans metaphysically. As such, as Jacob argues, Science created nature lawful.[351] Similarly, Bennett argued that the Enlightenment aimed to replace the notion of knowledge as mystical divine clues with a clear and certain scientific approach, and to move away from a view

---

[348] Ibid.,1.

[349] Ibid., 113.

[350] Jacob, *"Scientific Culture,"* 115.

[351] Ibid., 74.

of nature as sacred and instead utilize its resources for practical purposes.[352]

Additionally, people's trust in religion was also impacted by the secularization of the cosmos. All of the Christian tradition's principles would unavoidably be called into question if everything operated according to scientific principles because as the definition of creation evolved, the human understanding and perception of the Creator also underwent a transformation.[353] The mechanical theory of nature holds that all that is necessary for people to gain from nature is a basic understanding of science and mathematical principles. If a God existed, He may very well be aloof or, according to the physicist P.S. Laplace (1749–1837), a superfluous hypothesis.

It is not surprising that the religious beliefs of 18th-century engineers, inventors, and businesspeople had significantly decreased, if not disappeared entirely, over time. This is evident when we examine their ideals. However, it is noteworthy that their focus on advancing science and technology remained constant even as their faith declined. According to Jacob, this shift towards a secular worldview was a key aspect of the Enlightenment and played a critical role in the advancement of technology and subsequent economic development.[354]

The fall in religious belief, combined with an increase in curiosity in modern science, may be related to the different limits of modern technology. However, this situation was essentially related to how a progressively scientifically orientated civilization assessed truth and deception. The Age of Enlightenment marked a shift towards a more empirical and verifiable approach to discourse, where the rules and criteria for determining truth and effectiveness underwent a transformation.[355] To put it another way, the standards for truth become more reasonable and empirical. Graphs and algorithms could not illustrate the veracity of faith and intuition. In general, the widespread dissemination of scientific knowledge changed how people viewed the world, prompting them to endeavor to shape both their lives and the world around them to align with the newly acquired technological perspective, which was deemed to be both genuine and valuable. The

---

352 Bennett, "*Unthinking Faith and Enlightenment,*" 7.

353 Jacob, "*Scientific Culture*", 74.

354 Ibid., 127.

355 Mokyr, "The Intellectual Origins," 303.

secular attitude of the Enlightenment, according to Jacob, spurred people to pursue success in the world's life via technology and science and urged them to want to dominate and exploit the natural environment. Similarly, we might argue that scientific progress today is basically motivated by the influence that scientism or the scientific mindset has on any civilization.

The findings of Jacob and Mokyr provide more support to Nasr's claims that human behaviors that degrade the environment result from replacing religious viewpoints with the modern scientific viewpoint. What then should contemporary civilization do to seek a long-term solution to the ecological crisis, according to Nasr? Nasr's suggestions for all civilizations follow from this.

According to Nasr, adopting a sacred or religious view of nature is the key to resolving the environmental crisis. He suggests a number of related measures to do this. He claims that unless modern science and the scientific viewpoint are widely critiqued and its shortcomings are made clear, the sacred image of nature cannot be restored.[356] First and foremost, Nasr asserts that a modern human being must acknowledge the existence of God. The sacred image of the environment must therefore be restored, in accordance with conventional sources.[357] This would require reaffirming the belief that humans are endowed with superior faculties of awareness that extend beyond reason.[358] The religious perspective of the environment as a sign or a symbol of God can only be revitalized after accepting the truth of revelations and their sentience interpretation.[359]

According to Nasr, individuals must recognize the existence of metaphysical principles, such as the timeless principles we have discussed, that regulate both human ethics and the universe, in order to comprehend the interdependence between humanity and nature in the context of the Divine. This is the only way for humans to fully grasp the importance of ethical treatment towards the natural environment and fellow humans. Moreover, to prevent the harmful effects of modern science, scientific disciplines must be governed by all-encompassing metaphysical concepts.[360]

---

[356] Nasr, *"Religion and the Order of Nature"*, 152-53.

[357] Ibid., 222.

[358] Ibid., 185, 223.

[359] Ibid.

[360] Nasr, *"Man and Nature,"* 117.

Nasr's Views of Science and Technology

Nasr is not an outsider evaluating modern science and its history from a religious perspective. A highly educated individual, he pursued his academic interests in physics, mathematics, and geology as a student at MIT, and furthered his education by studying the history of science at Harvard University. In 1958, he successfully defended his doctoral book on the topic of "Conceptions of Nature in Islamic Thought," which was subsequently published as "An Introduction to Islamic Cosmological Doctrines." [361] In this work, Nasr highlights the possibility of a natural science grounded in metaphysics and ontology, as demonstrated by medieval Muslim scientists. He also argues that this kind of science can be a means of knowing God.[362] During the era when logical positivism and later the naturalism of John Dewey, W.V.O. Quine, and their adherents dominated the philosophy of science, Nasr demonstrated that the basic principles of the natural sciences stem from the prevailing worldview of the individuals who create them.[363] [364] He argued that these sciences cannot be fully understood by those with a different worldview, as seen in the failure of modern scientists and philosophers to appreciate Islamic sciences.[365] This idea, that worldview plays a role in the pursuit of science, was later popularized by Thomas Kuhn in his book *"The Structure of Scientific Revolutions,"* published in 1962.[366]

In 1966, Nasr delivered the Rockefeller lectures at the University of Chicago, where he identified the secularization of nature inherent in the modern scientific worldview as the root cause of the environmental crisis. In his book "Science and Civilization in Islam," Nasr expanded on this idea, emphasizing that the dominant worldview of society shapes the sciences it

---

361 Seyyed Hossein Nasr. "An Introduction to Islamic Cosmological Doctrines." Cambridge, Massachusetts: Harvard University Press, 1964.

362 Ibid., 277

363 Peter Godfrey-Smith. "Theory and Reality: An Introduction to the Philosophy of Science". Chicago: The University of Chicago Press, 2003, 237.

364 Alex Rosenberg. "Philosophy of Science: A Contemporary Introduction." 2nd edition. New York: Routledge, 2005, 81.198.

365 Nasr, *Introduction to Islamic Cosmological Doctrines*, xix.

366 Thomas S. Kuhn. "The Structure of Scientific Revolutions." 3rd edition. Chicago and London: The University of Chicago Press, 1996.

produces.[367] [368] Although Nasr and Kuhn share similar perspectives on the impact of worldviews in shaping scientific inquiry, they differ on several key points. For example, Kuhn rejects the notion of objective truth and the existence of transcendent reality, setting him apart from Nasr.[369] Kuhn and his followers, including Paul Feyerabend, argue that a paradigm or worldview is the product of historical and cultural influences, rather than unchanging, non-historical principles. They believe that knowledge is inherently tied to a specific paradigm or worldview, making it impossible to assess the truth of knowledge without considering its associated paradigm. Kuhnians argue that paradigms cannot be comprehensively understood from each other's standpoint because they are incompatible. This contradicts the empiricist belief that a new paradigm emerges only from scientific progress towards truth. Therefore, Kuhnians conclude that modern science is not intrinsically more objective than other human pursuits.[370]

The majority of scientists and philosophers of science, who are positivists and empiricists, believe that the human observer does not play a role in determining objective scientific knowledge. In contrast, proponents of Kuhnian relativism adopt a distinct position, asserting that all knowledge is the product of the perceiving self, which is shaped by historical and cultural factors. This perspective, as noted by Ibrahim Kalin, denies the existence of any objective reality in the natural world.[371] Both Kuhnians and their positivist/empiricist counterparts deny the existence of the Intellect and reduce the nature of the individual self to their senses and reason by rejecting any transcendent reality.

Nasr argues that the differentiation between traditional and modern science is established on the presence or absence of the Intellect and revelation. Therefore, he believes that the Islamic traditional sciences, which operate on principles comparable to perennial principles, cannot be under-

---

367 Seyyed Hossein Nasr, *"Science and Civilization in Islam"*, 2nd edition (Cambridge, UK: The Islamic Texts Society, 2007)

368 Nasr, "An Intellectual Autobiography," 30.

369 Godfrey-Smith, *"Theory and Reality,"* 96.

370 Paul Feyerabend, *"Against Method,"* 4th edition (London: Verso, 2010), 211-213.

371 Ibrahim Kalin, "Three Views of Science in the Islamic World," in *"God, Life and the Cosmos: Christian and Islamic Perspectives"*, eds. Ted Peters, Muzaffar Iqbal and Syed Nomanul Haq (Aldershot, UK: Ashgate, 2002), 58-59.

stood within the modern scientific paradigm that denies any transcendent reality and the existence of the Intellect beyond ordinary reason. However, the opposite is not true since the traditional paradigm's Intellect does not reject the validity of ordinary reason and its corresponding knowledge at the material level. Therefore, according to the traditional paradigm, there is no doctrine of incommensurability with modern science. The traditional paradigm acknowledges the appropriate position of modern scientific perspectives that are restricted to the material level. Nasr differs from both empiricists and Kuhnians by asserting that every religious tradition has immutable, non-historical principles of truth within its metaphysics. These principles can serve as a criterion for evaluating the objectivity of knowledge in any human endeavor, regardless of cultural norms or historical context. This standpoint starkly contrasts with the positivist and even the Kuhnian relativist viewpoints, which all adhere to the empiricist and rationalist method of knowledge that typifies contemporary science.[372]

According to Nasr, the manifold constraints of modern science are intertwined as they originate from its neglect of the "pearls of archetypes" or the essential reality of Forms that convey significance, in preference to their physical manifestations. This lack of emphasis on the heart or Intellect, which al-Ghazzali regarded as essential for comprehending the signs of God, accounts for such a disregard. Hence, Nasr categorizes the limitations evident in modern science as follows:

> Modern science operates under the implicit belief that matter is the source of all-natural existence, and any "forces" in nature must stem from matter. This reductionist approach, where complex things are understood by dissecting them into their most basic, material components, is considered a fundamental aspect of modern and postmodern thought, according to Nasr.[373] The revelations of Quantum physics suggest the potential for an undetectable genesis of the universe. Huston Smith notes that contemporary postmodern science is increasingly accepting and respectful of the

[372] Seyyed Hossein Nasr, *"Man and Nature: The Spiritual Crisis in Modern Man"* (Chicago: ABC International Group, 1997), 115

[373] Seyyed Hossein Nasr, "In the Beginning was Consciousness," in *"The Essential Seyyed Hossein Nasr"*, ed. William C. Chittick (Bloomington, IN: World Wisdom, 2007), 225.

> unseen.[374] Nasr is aware of the recent advancements in modern physics, which challenge the traditional view of matter as the foundation of the visible universe. Despite this, he cautions against the prevailing mindset: Even though modern quantum mechanics categorically denies the conventional perception of matter, there persists in the general domain a dependence on a materialistic outlook that ultimately reduces everything to mere "matter."[375] Additionally, modern science continues to seek out phenomena that can be perceived by the senses and understood through reason, driven by the requirement of demonstration.[376]

The emphasis on material causes stems from the rejection of the hylomorphism doctrine in rationalism. Hylomorphism posits that any physical entity is composed of both its "form," which encompasses its qualities and "matter," which represents pure potentiality. Mulla Sadra posits that the dog's classification as a dog is due to its animal form, rather than its particular matter. Likewise, the pig is considered a pig because of its form, not its matter, according to his philosophy.[377]

Nasr claimed that individuals who have been impacted by the scientific perspective are typically inclined to perceive the entirety as merely the sum of its constituent elements. Physicists continue to pursue the fundamental particles or building blocks of the universe, which the general public often envisions as tiny billiard balls that combine to form all existing entities. However, Nasr maintains that although the number "1" can be expressed as the sum of its fractions in mathematics, the essence of an entity cannot be reduced to the sum of its constituent parts.[378] Nasr notes that qualities don't accumulate as quantities do. To illustrate, a complete brick possesses qualitative properties that differ from those of two halves of brick, just as one-third of a flower does not exhibit the same qualities as one-third of the

---

374 Huston Smith, "The Ambiguity of Matter," *Sophia,* Vol. 3 No. 2 Winter, 1997, p. 20.

375 Nasr, "In the Beginning was Consciousness," 224.

376 Seyyed Hossein Nasr, "*Religion and the Order of Nature*" (New York: Oxford University Press, 1996), 152.

377 Sadr al-Din Shirazi, "*The Wisdom of the Throne: An Introduction to the Philosophy of Mulla Sadra*", trans. James Morris (Princeton: Princeton University Press, 1981), 146.

378 Ibid., 225. Also, see Wolfgang Smith, "The Plague of Scientistic Belief," in "*Science and the Myth of Progress*" (Bloomington, IN: World Wisdom, 2003), 223.

entire flower.[379] However, the dominant scientific community and even a significant portion of the public continue to follow the mechanistic model of classical physics, treating the whole as simply the sum of it. Contemporary science excels in analyzing quantitative components of wholes but fails to grasp the essence of the whole or make qualitative sense of it.[380] Modern science's limitation can be attributed to rationalism, which overlooks the "form" of entities.

According to Seyyed Hossein Nasr, modern science has reduced corporeal existence to pure quantity, ignoring all of its qualitative aspects. The consequence of this is that a perspective has been established in which life and consciousness are separated from pure inert matter.[381] Current technology can only perceive physical reality or material aspects and can only observe what is quantifiable. Quality, which is an essential aspect of existence, cannot be measured, and it can only be perceived directly. Modern science is limited to measuring physical manifestations that we have pre-determined to represent specific values, qualities, or meanings. While it can identify physical indicators of life or death based on our definitions, it is unable to distinguish between the two states themselves.[382] Regrettably, contemporary science is deficient in capturing the qualitative aspect of tangible reality and can merely provide insights into the physical world using quantitative measurements. Wolfgang Smith argues that since every physical entity possesses some value, quality, or significance, the physical entity recognized by modern science ought not to exist. In essence, modern science only offers a limited view of the world, ignoring the non-quantifiable aspects of reality that are essential to understanding the fullness of existence.[383]

According to Nasr, modern science portrays a universe that does not accommodate man's spirit, mind, or psyche, resulting in an "inhuman"

---

379 Seyyed Hossein Nasr, "*Knowledge and the Sacred*" (Albany: State University of New York, 1989), 236.

380 Bryan Appleyard, "*Understanding the Present: An Alternative History of Science*" (London: Tauris Parke Paperbacks, 2004), 191.

381 Nasr, "In the Beginning was Consciousness," 224.

382 Smith, "*Forgotten Truth*," 16.

383 Wolfgang Smith, "*Sophia Perennis* and Modern Science," in "*The Philosophy of Seyyed Hossein Nasr*, The Library of Living Philosophers," Volume XXVII, eds. Lewis Edwin Hahn, Randall E. Auxier, Lucian W. Stone, Jr. (Chicago: Open Court Publishing Company, 2001), 473.

universe that appears disconnected from the human condition.[384] Modern science doesn't consider the role of the human self or soul in perception in its investigations. It's as if our perception of the universe doesn't depend on what we believe or know. This occurs in two ways. Firstly, modern science is not primarily based on direct experiences of the world. It relies on data obtained through special conditions and instruments, rather than observable facts like the functioning of electricity, chemical reactions, or the laws of gravitation.[385] Secondly, modern science disregards the values and meanings that humans hold or perceive about the cosmos. For example, it would approach the determination of the reality of the Kaaba in Mecca in the same manner as any other physical object. Modern science is capable of quantifying the physical characteristics of objects such as a mountain's height, a fire's heat, or a lake's depth. However, it is incapable of providing commentary on qualitative aspects like beauty or meaning, which cannot be quantified. Such aspects are comprehended by the human intellect or heart, which transcends physical reality and grasps the significance of things.

Nasr argues that the success of applied science does not necessarily imply the infallibility of the scientific theories underlying it.[386] This position aligns with the philosophy of science's concept of "under determination." According to this theory, a set of data can be explained by multiple theories, which was a subject of discussion in the early 20th century. As a result, objectivity in modern science is limited to a certain set of assumptions and conditions, making it possible to attribute a specific cause to a particular observation. The notion of under-determination becomes more apparent when dealing with theories that incorporate intuitive knowledge.[387] For example, a patient can be treated for a particular illness using both traditional Islamic medicinal theories and those based on empirical evidence in modern medicine. Nasr believes that scientism is a failure to recognize these limitations of modern science. In the upcoming pages, we will delve into how scientism contradicts the timeless principles of Islam, as discussed

---

384 Seyyed Hossein Nasr, *"Traditional Islam in the Modern World,"* (London and New York: Kegan Paul International, 1994), 99.

385 Smith, *"Forgotten Truth,"* 98. While religion also claims that things are more than they appear to be, it promises knowledge by the 'eye of the heart' *within us*.

386 Nasr, *"Man and Nature"*, 115.

387 See Richard Dewitt, *"Worldviews: An Introduction to the History and Philosophy of Science"* (Oxford, UK: Blackwell, 2004), 48-49

in Islamic metaphysics, and how it can lead the mind and soul astray from the concept of unity that embodies *tawhid*. This will help us gain a better understanding of the influence of scientism on Islamic values.

Critique to Scientism

Scientism, according to Nasr, is the belief that science is the only valid way of acquiring knowledge about the world, and that all other forms of knowledge are irrelevant or illegitimate. This view, he argues, denies the fundamental Islamic principle of *tawhid*, or the unity of God and the universe, and the hierarchy of reality. Nasr explains, "Scientism claims that there is no reality other than that which can be observed by the physical senses and measured by scientific instruments. Thus, it denies the hierarchical structure of reality that is at the heart of Islamic philosophy".[388] Nasr also argues that scientism leads to a reductionist and materialist worldview that ignores the spiritual dimensions of existence. This, in turn, leads to a distorted understanding of the world and a neglect of important moral and spiritual values. He writes, "Scientism reduces everything to its physical components and ignores the spiritual dimensions of existence. This leads to a distorted understanding of the world and a neglect of important moral and spiritual values".[389]

One area where this reductionism is particularly evident is in the theory of evolution. Nasr argues that the theory of evolution is based on a flawed understanding of the nature of reality, which assumes that everything can be reduced to its physical components and that there is no spiritual dimension to existence. He explains, "The theory of evolution is based on the assumption that all living beings can be reduced to their material components and that there is no spiritual dimension to existence. This is a fundamental flaw in the theory, which leads to a reductionist and materialist worldview".[390] Nasr also points out that the theory of evolution has been used to justify a range of immoral and destructive practices, such as eugenics and social Darwinism. He argues that this is because the reductionist and materialist worldview of scientism fails to recognize the complexity and richness of

---

[388] Nasr, S. H. (2014). Islamic philosophy from its origin to the present: Philosophy in the land of prophecy. SUNY Press.

[389] Ibid. p, 12.

[390] Ibid. p, 13.

the natural world and reduces everything to a set of mechanistic processes.

Nasr argues that modern science has become a way of *takthir*, which means "reductionism" in Arabic. Modern science reduces everything to its component parts and does not recognize the interdependence and interconnectedness of all things. This reductionism has led to the destruction of the environment and the ecological crisis that we are facing today. Nasr claims that "modern science is an expression of a materialistic worldview that has lost touch with the spiritual dimension of life".[391] Without an understanding of the interconnectedness of all things, modern science will continue to contribute to the destruction of the environment.

Nasr's critique of scientism highlights the dangers of reducing reality to a single dimension and ignoring the spiritual dimensions of existence. His arguments against the theory of evolution and his critique of modern science as a way of *takthir* remind us of the need to approach the world with humility and respect for its complexity and diversity and to recognize the importance of *tawhid* and the hierarchy of reality.

Scientism Rejects the Concept of Tawhid and the Idea of a Hierarchical Order in Reality

The cosmology of Ibn Sina, which is peripatetic in nature, consists of a system of Intellects or Archangels arranged in a hierarchical structure. Each of these Intellects oversees a lower sphere, which symbolizes a framework of knowledge and consciousness. This concept was further expounded by Ibn 'Arabi, who established a hierarchy based on the degree of divine attributes manifested in various entities. The more elevated a reality, the more it is imbued with Absolute Being or Awareness. Islamic philosophers, such as Nasr, contend that being cannot be dissociated from the knowledge and, thus, awareness. They perceive the cosmic level of existence as a series of levels of knowledge and awareness.[392] Islamic metaphysics holds that the unity of reality is essential to understand the relationship between various levels of existence and awareness of God's qualities. Islamic cosmology portrays the relationship between God and His infinite creation as a hier-

[391] Ibid. p, 54.

[392] Nasr, "In the Beginning was Consciousness," 223.

archy. In Islam, the concept of *tawhid*, which is the belief in one God, serves as the foundation for all metaphysics.

The Islamic belief in a God who pervades all levels of reality with both Being and Awareness contradicts the assumption in modern science that matters, in the conventional Newtonian perspective, is the foundation of all matters. This denial of the unity and hierarchy of reality also leads to the view that God's existence is unnecessary in the scientific field. By focusing solely on physical reality, modern science makes the concept of God irrelevant.[393] According to Nasr's metaphysics, by promoting modern science, modernist and fundamentalist Muslims are essentially allowing science to define the cosmos solely in terms of material reality. This leads to a situation where the physical world is perceived as an independent entity that can be fully understood and studied without considering a higher reality. As a result, the believer's view of the world becomes limited to the material realm.[394] Despite their belief in God, modernist and fundamentalist Muslims, with their embrace of scientism, miss the understanding of God as the all-pervading reality. The reason for this is that understanding God as the all-encompassing reality necessitates acknowledging the various levels of reality within the indivisible reality of God.[395]

Scientism, by prioritizing material or physical forces as the foundation of all phenomena, denies the hierarchy of reality and restricts the belief in God as an absolute "other" beyond the natural world. This results in a decrease in the perception of God as an only transcendent reality and may lead to a complete abandonment of belief in God. The universe is observed as a multiplicity of entities with no connection to each other or to God, and the relationship between humans and the cosmos as well as the microcosmic-macrocosmic relationship is ignored. The natural order is seen as mere signs of God's kindness and generosity for human benefit rather than as a means of higher knowledge of God and oneself. Thus, scientism undermines the contemplation of *tawhid* and the unity it represents.

---

[393] Seyyed Hossein Nasr, "Islamic Worldview and Modern Science, *Islamic Thought and Scientific Creativity*," Vol. 7 (1), 1996, pp.11-12.

[394] Seyyed Hossein Nasr, *"A Young Muslim's Guide to the Modern World,"* 2nd edition (Chicago: Kazi Publications, 1994), 182.

[395] Seyyed Hossein Nasr, *"The Need for a Sacred Science,"* (New York: State University of New York, 1993),11.

Following the hierarchical structure of reality, where all levels of existence reflect unchanging prototypes in God's wisdom, the significance of any entity ultimately lies solely in God. Nasr argues that without accepting the ultimate reality in a metaphysical sense, there can be no ultimate meaning.[396] Because entities can only have ultimate realities in one level of existence, recognizing *tawhid* is essential in understanding the ultimate purpose of the natural world. However, contemporary science, which is limited to providing knowledge of physical reality, cannot contribute to the understanding of the essential significance of the natural world or *tawhid*. Instead, by disregarding the hierarchy of reality, modern science focuses on the multitude of physical entities and detracts from a sense of unity.

In theory, what is the method for comprehending the ultimate truths of natural entities as unchanging archetypes in God? Prior chapters have explored how traditional sciences aimed to portray a sense of coherence in the universe, emphasizing the interrelatedness of all things that stem from a single divine source. This concept of unity serves as a means for Muslims to reflect upon *tawhid*, or the fundamental oneness of all reality.

Nasr posits that traditional Islamic sciences included an internal method for comprehending the external world by purifying one's heart or inner self.[397] As a Sufi, al-Ghazzali held that once the heart is purified through the Sufi path, the knowledge of the authentic nature of things naturally manifests within the heart.[398] According to Nasr, traditional scientists sought to understand the true nature of things with pure hearts, inspired by a prayer of Prophet Muhammad.[399] Their ultimate goal was to comprehend the reality of all things in relation to God. Nasr emphasizes that 'Umar Khayyam (1048-1131), a distinguished mathematician and Sufi poet, recog-

[396] Nasr, "In the Beginning was Consciousness,"227. "Values and meaning cannot be volatile, they must endure. If nothing in science endures, it can offer no meaning." Appleyard, *Understanding the Present*, 195.

[397] Seyyed Hossein Nasr, "Reflections on Methodology in the Islamic Sciences," *Hamdard Islamicus*, Vol. III No.3, (1980), 3-13.

[398] See Abu Hamid al-Ghazzali, "The Elaboration of the Marvels of the Heart"," in "*Knowledge of God in Classical Sufism,*" trans. John Renard (New York: Paulist Press, 2004), 301.

[399] Nasr, "*Science and Civilization in Islam,*" 352-353. Nasr observes that the famous prayer by the Prophet Mu_ammad, " O Lord show us things as they really are," set the course for Islamic quest for knowledge at the deepest level for all entities in the cosmos.

nized the adoption of an inner approach to knowledge by scientists, who were frequently Sufis. He held the belief that this approach was superior to a purely rational one:

> The Sufis pursue knowledge through the purification of their inner being and the refinement of their dispositions, rather than through meditation or discursive thinking. They rid the rational soul of the impurities of nature and the body until it is transformed into a pure substance. Then, the soul is able to directly perceive the spiritual world, without any doubt or ambiguity, and the forms of that world are accurately reflected in it. This path is the best way to attain perfection, as it removes all obstacles and veils that prevent one from accessing the perfections of God. Whatever ignorance arises within humans is due to the impurities of their nature. However, when these impurities are eliminated, the truth of things becomes evident. The Prophet Muhammad – peace be upon him – indicated that during one's lifetime, inspirations from God are present and that the Gnostics follow their intuition, rather than relying on discursive thought.[400]

The Sufi approach to understanding reality, known as *tahqiq*,[401] involves gaining knowledge through the heart, and eliminating all doubts or uncertainties. According to Khayyam, it is necessary to purify the rational mind to attain the "eye of the heart" and allow the intellect to grasp the "forms" of the spiritual world. He believed that this method was the most effective way to understand the essential nature of things. Furthermore, it is worth noting that Khayyam, a renowned Persian mathematician, astronomer, and philosopher, was a proponent of the concept of tawhid, which emphasizes the unity and oneness of reality. He held the view that there was no distinction between the knowledge of the "perfections of God" and the truth of things as they are, indicating a profound connection between the divine and the material world. However, Khayyam and other Sufi scientists, including Jabir Ibn Hayyan, Nasir al-Din al-Tusi, Qutb al-Din Shirazi, and Baha' al-Din 'Amili, did not abandon their rational and empirical

---

[400] 'Umar Khayyam quoted in Nasr, "*Science and Civilization in Islam*," 34.

[401] See William C. Chittick, "*Science of the Cosmos, Science of the Soul: The Pertinence of Islamic Cosmology in the Modern World*" (Oxford, UK: Oneworld Publication, 2007), 23-24.

methods of inquiry, despite embracing the inner approach to knowledge. They recognized that the pursuit of knowledge required a balance between the spiritual and the physical aspects of reality. Moreover, these scientists believed in a hierarchy of reality for all entities, from the lowest to the highest, emphasizing the interconnectedness of all things. They recognized that each level of reality had its unique characteristics and properties and that a complete understanding of the universe required an appreciation of this hierarchy. In essence, Khayyam and other Sufi scientists held that the pursuit of knowledge was a multifaceted and holistic endeavor that required the integration of both the spiritual and the material aspects of reality. They believed that by recognizing the unity of reality and appreciating the hierarchy of entities, they could gain a deeper understanding of the universe and its place within it.[402] The study of the nature of things in traditional Islamic science incorporated a range of methodologies, including rational, empirical, self-purification through illumination, and revelation. This diverse approach was driven by the belief in a hierarchy of reality that emphasized the interconnectedness of all entities.[403] Conversely, contemporary science, which limits itself to rational and empirical methods, falls short of offering a complete comprehension of the fundamental truths about natural entities. Chittick explains that modern science largely relies on "hearsay" and lacks the insights that can be gained through the inner approach to knowledge (*tahqiq*).[404]

Many scientists lack the required skills and equipment to independently verify all of the theories they accept as true. Their confidence is based on the agreement of previous scientists, who themselves relied on the findings of others. Even those who have tested theories in their labs do not have complete certainty in their hearts:

> It can be argued that a modern scientist who makes a new discovery may consider it personally "verified" and "realized." Nonetheless, the Muslim intellectual tradition would not consider this as tahqiq since it fails to delve deep enough into the essence of things to fully

[402] Nasr, "Reflections on Methodology in the Islamic Sciences,"7-10.

[403] Ibid., 8. Also, see Osman Bakar, "Reformulating A Comprehensive Relationship Between Religion and Science: An Islamic Perspective," *Islam and Science,"* Vol. 1 (June 2003) No. 1, pp. 35-36.

[404] Chittick, *"Science of the Cosmos,"* 24.

recognize their true nature at the soul and spirit level.[405]

As an example, some religious leaders like Prophets and saints have claimed to have a deep understanding and realization of the concept of the unity of reality (tawhid) in their hearts. However, even scientists cannot make similar claims about their understanding of theories like Newton's theory of gravitation, as the proof of such theories is always based on external evidence, not a personal realization at the core of their being. Unlike modern sciences, traditional sciences and the Sufi path offer a holistic view of reality. Traditional sciences acknowledge the higher realities of physical entities, while the Sufi path aims to understand these realities at their core through a connection with God. Both traditional sciences and the Sufi path serve as a retreat from the fragmented perception of the world, towards a unified understanding of reality.

The topic of scientism's denial of the principles of reality's unity, hierarchy, and purpose in metaphysics has been addressed. We shall now investigate how scientism can impact the human psyche, along with providing instances. The limitations of modern science are reflected in scientific knowledge. If we allow it to shape our perception of reality, we run the risk of falling into the traps of scientism. Nasr uses this scenario in contemporary society to explain how young Muslim students in traditional madrassahs did not abandon their prayer rituals even after reading Khayyam's algebra or Jabir Ibn Hayyan's alchemical writings, unlike many current students who lose their religious grounding upon studying modern mathematics and chemistry.[406] As previously discussed, to fulfill our role as God's representatives on earth, it is necessary to have a thorough understanding of all things. This allows us to approach them with proper awareness. The way we react to an entity is contingent on our awareness of it. If this awareness originates from God, who is the foundation of reality (*haqq*), then we possess the tools to gain a more profound comprehension of the entity's authentic essence.

The Quran and individuals with awareness of higher planes of existence,

[405] Ibid.

[406] Seyyed Hossein Nasr, "Islam and Modern Science," in "*Islam and Contemporary Society*," ed. Salem Azzam
(London: Longman Group, 1982), 179.

known as gnostics, characterize the earth as a "resting place" (40:64), the sky as a "well-guarded canopy" (21:32), and water as a divine bestowal from the heavens (2:22). This understanding, regardless of scientific explanations, allows us to have a deeper connection to these entities. These depictions not only provide relatable connotations but also express God's affection and attention towards us. This customary approach to understanding the universe grants significance and intention to everything, including our individual existences.

The modern scientific worldview considers the sky to be simply a collection of gaseous particles reflecting the color blue, stars as distant gaseous bodies emitting light from nuclear fusion, and water as the simple chemical formula H2O, consisting of two hydrogen atoms and one oxygen atom. This view openly conflicts with religious interpretations of reality. Scientific understanding of the stars, sky, and water as mere physical entities fails to capture the spiritual connections our souls may have with them. It provides no insight into the presence of a higher power or how to connect with it. The mere classification of water as H2O does not convey its potential for meaningful relationships with humans, the divine, or other entities in the universe.[407] The replacement of traditional knowledge with scientific knowledge results in the loss of a way for humans to connect with a higher power. Moreover, scientific comprehension reinforces the conviction that the physical realm is simply a series of occurrences lacking any inherent link to humans or a divine being.

According to Nasr, scientific theories that contradict fundamental religious beliefs about the origins and destiny of life and the cosmos have significant implications. Scientists construct theories about the cosmos' origins, age, and eventual fate based on physical observations. As noted by Nasr, the speculative character of contemporary science is emphasized by the existence of diverse theories, including the Multi-verse theory, String theory, and Big Bang theory.[408] Additionally, scientific theories reject the religious or spiritual perspectives on the beginning and end of the universe. Nasr notes that even if a theory posits a divine entity as the originator, it reduces the relationship between the divine and the cosmos to simply physical

[407] Chittick, "*Science of the Cosmos,*" 100.

[408] Seyyed Hossein Nasr, "The Question of Cosmogenesis – The Cosmos as a Subject of Scientific Study," "*Islam and Science,* Vol. 4"(Summer 2006) No.1, p. 53

terms, and fails to acknowledge the distinction between the divine and material realms. This separation of God from His creation is evident in scientific cosmogenesis theories, even those which acknowledge God as the creator. [409] [410] Scientific theories differ from religious or spiritual beliefs in their failure to reconcile the end of the universe with a divine origin at a higher level of being. Instead, they simply propose a "final death" for everything, leaving the universe without an ultimate goal. To put it briefly, contemporary scientific theories regarding the origin and ultimate fate of the universe challenge the fundamental principles of Islam, thus eroding the very foundation of Islamic belief.[411] [412]

Nasr doesn't support completely disregarding modern science's findings. Rather, he opposes scientism, which overlooks the limits of scientific knowledge. According to him, the religious and traditional scientific portrayal of nature, which is symbolic in nature, should not be disregarded in light of scientific knowledge. He believes that stating the sun is made entirely of incandescent gas is an incomplete depiction of its reality. It is equally valid to assert that the sun embodies the intelligible principle of the Universe, and this aspect is just as much a part of its existence as the physical properties that have been discovered through modern astronomy.[413]

While modern science and theories concerning reality may reject ideas such as *tawhid,* the hierarchy of reality, and the notion of ultimate meaning or purpose in the universe, they are primarily grounded in empirical evidence, although not entirely devoid of assumptions, at the material level. Regarding the theory of evolution, Nasr perceives it as largely reliant on conjectural concepts rather than concrete observations on the material plane. Nasr believes that it's crucial for Muslims to challenge the theory of evolution. This is because of its widespread influence in the modern world and its rejection of spiritual beliefs about the origin and purpose of life. [414] Nasr argues against the theory of evolution from scientific, metaphysical,

[409] Ibid., 54.

[410] Ibid.

[411] Ibid.

[412] Ibid.

[413] Nasr, "*Man and Nature,*" 120.

[414] Nasr, "*Traditional Islam in the Modern World,*" 105.

philosophical, logical, and religious perspectives.[415] We'll outline his objections to the theory from these various viewpoints.

Nasr Rejected the Theory of Evolution

Nasr explains that the Quranic view of God identifies Him as The One (*Al-Ahad*), The Living (*Al-Hayy*), The Giver of Life (*Al-Muhyi*), and The Creator (*Al-Khaliq*). In Islam, the concept that life originated as an accident is rejected because, according to Islamic belief, only the Source of existence (God) has the power to give life.[416] Nasr argues that belief in theistic evolution, which posits that God created species and then allowed them to evolve naturally, contradicts the Islamic understanding of God as "*Al-Muhyi*" and "*Al-Khaliq*".[417] Contemporary science is unable to either prove or disprove the role of God in the creation and is also incapable of causing the non-existence of any entity. Additionally, the theory of evolution conflicts with Islamic beliefs by rejecting consciousness as the origin of physical reality and regarding life and consciousness as mere byproducts of material factors. Nasr supports the concept of an unchanging species and argues that if all beings are constantly evolving from one to another species, as exemplified by an ant, then what we currently identify as an ant is simply a transitory stage in that evolutionary process. This poses a problem for the idea of a permanent reality of the ant species, as asserted by the Quran if the ant is continuously evolving.[418] [419]

Nasr challenges the scientific evidence for evolution. He cites biologists who argue that species appear abruptly in the paleontological record, contradicting the idea of gradual evolution from previous species.[420] The theory proposed by Darwin in the 19th century lacked substantial scientific evidence and was instead accepted because it aligned with the phil-

---

[415] Ibid., 106.

[416] Seyyed Hossein Nasr, "On Biological Origins," interview by Muzaffar Iqbal, in *"Islam, Science, Muslims, and Technology: Seyyed Hossein Nasr in Conversation with Muzaffar Iqbal"* (Alberta, Canada: Al-Qalam Publishing, 2007), 153.

[417] Ibid., 152.

[418] Ibid., 152.

[419] Nasr, "In the Beginning was Consciousness," 225.

[420] Ibid., 155.

osophical concept of advancement in the physical realm and fulfilled the materialists' desire for causation in a world without God.[421] [422] As Darwin acknowledged, he depended on the conjecture of imperceptible intermediary life forms to explain the observable biological data.[423] However, these transitional forms have never been found, which weakens his theory.[424] Moreover, the idea proposed by Darwin that the fundamental unit of life was a straightforward single cell, forming the basis for the theory that sophisticated organisms evolved from primitive cells over the ages, has been proven completely inaccurate.[425]

The theory of evolution only addresses the physical aspect of living entities and doesn't consider the non-physical aspect.[426] However, according to Nasr, Islam views a living entity as more than just its physicality.[427] The similarity in DNA between animals and humans doesn't necessarily prove evolution from one species to another.[428]

According to renowned scientists in the traditional Muslim world, such as Ibn Sina, the fact that certain characteristics are repeated in more complex species does not necessarily indicate biological evolution. Instead, it suggests that advanced life forms have inherited the abilities of simpler creatures and beyond.[429]

Nasr's argument suggests that for species "A" to evolve into species "B", there must be a pre-existing potential for species "B" within species "A".

---

421 Osman Bakar, "The Nature and Extent of Criticism of Evolutionary Theory," in "*Science and the Myth of Progress*" (Bloomington, IN: World Wisdom, 2003), 174.

422 Frithjof Schuon, "*Form and Substance in the Religions*" (Bloomington, IN: World Wisdom, 2002), 65, quoted in Bakar, "The Nature and Extent of Criticism of Evolutionary Theory," 169.

423 Stephen Meyer, "*Signature in the Cell: DNA and the evidence for Intelligent Design*" (New York: HarperCollins, 2009), 424; Nasr, "*Knowledge and the Scared*", 237,239.

424 David Berlinski, "*The Devil's Delusion: Atheism and Its Scientific Pretensions*" (New York: Crown Forum, 2008), 188-193.

425 Meyer, "*Signature in the Cell*", 44.

426 Bakar, "The Nature and Extent of Criticism of Evolutionary Theory," 174-179

427 Nasr, "On Biological Origins," 162-63.

428 Ibid.

429 Nasr, "On Biological Origins," 163. In "Ibn Sina's cosmology for instance, the rational soul of the human being subsumes the animal soul and the vegetable soul." See Nasr, "*Introduction to Islamic Cosmological Doctrines*", 249-252.

This is because, according to logic, something cannot become another thing unless that other thing already exists in some way within it. Additionally, if species "A" possesses something greater than species "B", then species "B" cannot emerge from species "A". [430] Ignoring this logical gap, according to Nasr, would require ignoring the existence of the higher. The notion at the core of the metaphysical argument is that the higher reality cannot emerge from the lower unless the potential of the higher is already inherent in the lower. This principle is exemplified in the human mind. In metaphysics, physical forms are considered weaker versions of their unchanging counterparts in higher planes of existence, creating a hierarchy of reality.[431] Reality flows downward from its more permanent source above, not the other way around. This means that physical entities on the earthly plane are limited to the parameters of their unchanging archetype and therefore, while small changes (micro-evolution) are possible, significant transformations such as shifting from one to another species (macro-evolution) are not:

> While micro-evolution can occur within a species' archetype, macro-evolution is not feasible. The archetype sets the boundaries within which micro-evolution can happen. A species has a certain range or width that encompasses the traits of all individuals within that species. Hence, different individuals can emerge within the same species, exhibiting distinct characteristics and adapting to environmental circumstances, without the species transforming into a new one.[432]

Nasr contends that the theory of evolution is not in line with religious beliefs, lacks scientific evidence, and goes against logical and metaphysical principles. According to him, the scientific "theory of evolution" conflicts with the religious belief that everything originates from fixed prototypes in God's understanding and will eventually return to Him. Instead, the theory of evolution presents a view that life started inexplicably from matter, evolved and multiplied randomly over time without an ultimate goal or significance. Thus, the metaphysical or religious viewpoint offers a perspective of *tawhid*, whereas the "theory of evolution", which is a

---

[430] Nasr, *"Knowledge and the Sacred"*, 237.

[431] Bakar, "The Nature and Extent of Criticism of Evolutionary Theory," 169.

[432] Nasr, "On Biological Origins," 154.

fundamental element of the contemporary scientific outlook, presents a viewpoint of diversity without any ultimate purpose or significance.

According to Nasr, rejecting the sacred origin and ultimate goal of living beings is likely to lead to destructive outcomes.[433] Similarly, philosopher George Grant shares Nasr's belief: The period during which Darwin's theories have held significant influence in the life sciences coincides with the time in which humans have caused the highest number of extinctions compared to any other time in history. One could argue that this shared historical context is a result of a common technological fate that led to both Darwin's scientific discoveries and humanity's domination of the natural world at a deeper level.[434] It should be noted that not all environmentalists condemn contemporary science. In fact, there are those who draw motivation from the theory of evolution, which Nasr does not endorse. This leads to questions about Nasr's claim that religious beliefs or the lack thereof shape human behavior towards the environment. These environmentalists can be broadly categorized as evolutionists who openly adopt Darwin's theory and pragmatists who indirectly support it. Thus, it's crucial to comprehend their perspectives and consider Nasr's response.

Bron Taylor explained the evolutionists' criticisms of religious worldviews in the "*Encyclopedia of Religion and Nature*", which edited,[435] also appears to support "nature-oriented spirituality"[436] or "dark green religion,"[437] referring to religions centered around direct human experiences with nature. Therefore, he is unhesitant in promoting new religions rooted in the Darwinian "evolutionary worldview." [438] Taylor contends that the notion that religious worldviews, particularly monotheistic religions like Christianity and Islam, have a beneficial effect on human behavior towards the environment is an idealistic belief that lacks evidence. However, the

---

433 Nasr, "*Religion and the Order of Nature*", 216.

434 George Grant, "*Technology and Justice*" (Concord, Ontario: Anansi Press, 1986), 65.

435 "*The Encyclopedia of Religion and Nature*," ed. Bron Taylor (New York: Thoemmes Continuum, 2005). See Taylor "Religious Studies and Environmental Concern" in "*The Encylopedia of Religion and Nature*," 1375.

436 Bron Taylor, "Religious Studies and Environmental Concern," 1375-76.

437 See Bron Taylor, "*Dark Green Religion: Nature, Spirituality and the Planetary Future*" (`Berkley, Los Angeles, London: University of California Press, 2010), Chapter 2.

438 Bron Taylor, "Introduction to *The Encyclopedia of Religion and Nature*, ed. Bron Taylor" (New York: Thoemmes Continuum, 2005), xx.

success of environmental activism in the Muslim world over the past two decades, driven by Islamic beliefs, as well as the growing environmental activism within Christian communities, clearly demonstrates the impact of religious perspectives. [439] [440] Taylor dismisses the impact of conventional religious beliefs on the environment and concentrates on contemporary ethical principles that have arisen after the advent of Darwinism. These principles, which can be derived from an evolutionary perspective, emphasize the shared ancestry and survival struggle of all life, fostering a sense of kinship. They shift humans from a privileged position of moral importance and instead acknowledge the interdependence of all living beings. [441]

Taylor's concept of "dark green religion" is characterized by a sense of unity with nature, but this feeling is not evidence for the theory of evolution. [442] [443] Instead, mainstream religious figures have repeatedly affirmed the unity of all things through personal experiences and have based their ethical teachings on this metaphysical understanding. These teachings have been supported for centuries.[444] Nasr's traditional views of religion, expressed in modern language, recognize the deep connection between humans and nature rooted in their common origin in God, surpassing Taylor's notion of a "common ancestor." Mainstream religions, particularly their spiritual aspects, are focused on God and emphasize following divine will. As a result, humans are not the primary focus of moral consideration, contrary to Taylor's perspective.

Ben Minteer and James Collins, renowned environmental pragmatism philosophers, recognize that ontology and metaphysics play a crucial role in establishing the intrinsic value of nature through environmental value theory.[445] However, their stance on the usefulness of metaphysics is ques-

[439] Taylor, "Religious Studies and Environmental Concern," 1376.

[440] Willis Jenkins, "After Lynn White: Religious Ethics and Environmental Problems" *Journal of Religious Ethics*, 37.2, 2009, p.296.

[441] Taylor, "Introduction," xx.

[442] Taylor, "*Dark Green Religion,*" Chapter 2: 13-41.

[443] Ibid., 23.

[444] "*Seeing God Everywhere: Essays on Nature and the Sacred,*" ed. Barry McDonald (Bloomington, Indiana: World Wisdom, 2003).

[445] Ben A. Minteer and James P. Collins "From Environmental to Ecological Ethics: Toward a Practical Ethics for Ecologists and Conservationists" *Sci Eng Ethics*(2008) 14:483-501 and 489.

tionable as they have been unable to address critiques from colleagues.[446] Examining the perspectives of pragmatist philosophers, it becomes clear that they believe the intrinsic value of nature has a naturalistic empirical basis rather than being rooted in traditional metaphysical beliefs. As an example, Minteer embraces John Dewey's natural piety, which was influenced by Darwin's evolution theory, as the quintessential mindset for environmental pragmatism.[447] [448] Bryan Norton, another well-known pragmatist, acknowledges the influence of Aldo Leopold, a pioneering non-religious environmentalist who accepted evolution as a factual occurrence. Norton, like Leopold, advocates for a naturalistic comprehension of environmental values supported by an epistemology that aligns with this perspective.[449] [450] According to Nasr, both evolutionists and pragmatists dismiss the innate spiritual nature of the natural world. While their focus on the physical realm may result in positive outcomes in the short term, they lack a long-term vision of the ultimate purpose and significance of nature. Consequently, their sole emphasis on the material world and support for evolution results in a fragmented and materialistic view of reality, potentially leading to negative implications for preserving the natural universe. Shaya Isenberg and Gene Thursby have pointed out that certain environmentalists critique mystical perspectives on the environmental crisis as impractical and detached from reality. However, these critics have yet to address the assertion made by perennial philosophers that all forms of activism rest on metaphysical assumptions, and flawed assumptions can result in ineffective activism.[451]

---

446 Ben A. Minteer "The Future of Environmental Philosophy" *Ethics & The Environment*, 12(2) 2007), 132. Also, see Bryan Norton "The Past and Future of Environmental Ethics/Philosophy" *Ethics & Environment*, 12(2) 2007,135.

447 John Dewey, "The Influence of Darwinism on Philosophy," in "*The Essential Dewey: Pragmatism, Education, Democracy*" Vol. 1(Bloomington, IN: Indiana University Press, 1998), 39-45.

448 Ben A. Minteer, "Pragmatism, Piety, and Environmental Ethics" *Worldviews* 1 2(2008) 179-196.

449 Aldo Leopold, "*A Sand County Almanac*" (New York: Ballantine Books, 1970), 116-117.

450 Norton, "The Past and Future of Environmental Ethics/Philosophy," 136; Also, see Bryan Norton, "Beyond Positivist Ecology: Toward and Integrated Ecological Ethics," *Sci Eng Ethics* (2008) 14:581-592, 582.

451 Shaya Isenberg and Gene Thursby, "Perennial Philosophy," in "*The Encyclopedia of Religion and Nature*", ed. Bron Taylor (New York: Thoemmes Continuum, 2005),1272.

Contemporary Science as a Means of Domination (Takthir)

According to Nasr, the shift from traditional to contemporary technology represents a shift towards pure diversity, resulting in a deviation from a comprehensive comprehension of Reality or Truth in its universal form, ultimately impacting *tawhid*.[452] In Islamic metaphysical terms, this shift in consciousness from *tawhid* to *takthir* refers to moving from perceiving unity to perceiving fragmentation and disconnection in our comprehension of the world. Traditional natural knowledge places emphasis on *tawhid* as the central perspective and objective, whereas modern science, confined to the physical realm, lacks a comparable unifying vision.[453]

Modern scientific knowledge contradicts timeless principles and instead provides a fragmented perspective, focusing on quantity rather than the holistic quality of entities. As a result, modern science's pursuit of quantitative analysis leads to further division and specialization. In short, *takthir* is the only outcome that modern science can offer.

As previously mentioned, higher truths and unity in metaphysics cannot be understood through ordinary reasoning, but rather through a higher reality within us, the brain or heart. In the 17th century, Mulla Sadra cautioned against the hazards of *takthir* that may arise if we neglect the use of our intellect and solely rely on our senses. He stated, "It is the nature of the intellect to unify the many, while the senses divide the one." [454] Today, the dominance of empiricism and the limited rationality of the sensory world in contemporary science has validated Sadra's warning. By embracing the limitations of contemporary science, scientism hinders our contemplation of *tawhid*, thereby depriving us of understanding the ultimate purpose of our scientific pursuits. The solution is not to rely solely on contemporary science, but to approach scientific inquiry with a unifying perspective. As Nasr stated, "Fragmented knowledge can only be integrated into a holistic understanding when there is a pre-existing intellectual vision of unity." [455]

[452] Nasr, "*Need for a Sacred Science*", 80-81.

[453] Chittick, "*Science of the Cosmos*", 12.

[454] Mulla Sadra, I, 3,380 quoted in Ibrahim Kalin, "Mulla Sadra's Realist Ontology of the Intelligibles and Theory of Knowledge," *The Muslim World*, Vol. 91, January 2004, 99.

[455] Nasr, "*Islam and the Plight of Modern Man*", 7.

This new approach would need trust in the unity of reality as revealed in religious scripture or the wisdom of the sages.[456]

The extent to which scientism appears convincing directs a person towards the viewpoint of *'takthir'*. In comparison to traditional times when contemporary science was not present, those swayed by scientism view their perceptual abilities as limited to basic reasoning. Furthermore, scientism promotes the idea that the cosmos lacks any connection to a divine presence, even if it exists, and holds no importance besides serving materialistic wants and needs. The perspective of unity, where everything in the universe is interconnected, is disregarded by modern science, resulting in a self-centered and materialistic mindset when influenced by scientism.[457]

Consequently, individuals with a rationalistic and scientistic outlook, who identify as Promethean humans, disregard religious or metaphysical convictions about the natural world and believe that they have the liberty to modify the environment based solely on their own reasoning and self-interest. [458] Nasr believes that the change in the human-nature relationship, from a traditional viewpoint where everything in nature held sacred significance to a scientistic view where nature is solely a source of worldly needs, is the main reason for humanity's destructive approach towards the environment. A Muslim who is swayed by scientism may still hold onto their belief in a divine deity and the *Shari'a,* but their thoughts can eventually be impacted by scientific perspectives. For example, they can't remain untouched by the manner in which modern science presents theories like the creation of the universe or the theory of evolution. According to Nasr's warning, it is essential to recognize the impact of secular science that contradicts the Qur'anic perspective on the creation of the world, even for individuals who are religious. Disregarding such fundamental issues, which may initially seem more theoretical than immediate, by claiming that society has more pressing concerns, would be unwise.[459]

---

456 Martin Lings, "*Ancient Beliefs and Modern Superstitions,*" 2nd edition (London and Boston: Unwin Paperbacks,1980),36-37.

457 See Eric Schliesser, "Wonder in the Face of Scientific Revolutions: Adam Smith on Newton's 'Proof' of Copernicanism," *British Journal for the History of Philosophy* 13(4) 2005, pp. 697-732.

458 Nasr, "*Knowledge and the Sacred*", 161.

459 Nasr, "Islam and the Problem of Modern Science,"136.

Even if Muslim rejects scientific theories solely based on their faith, they cannot escape the influence of the scientific worldview due to the modern world they live in. The essence of modern science is to categorize and quantify everything, which does not contribute to the realization of *'tawhid'* but instead leads to *'takthir'* in every field it is applied. The scientific approach encourages an unending pursuit of objectives restricted to the physical or psychological realm, which can be comprehended using ordinary reasoning and senses.[460]

We are not solely addressing the expansion of scientific possibilities in the physical sciences, which concentrate on external factors and are constrained to the material realm. We are also referring to the human and social sciences, which aim to imitate the methods of the material sciences by pursuing maximum precision and quantification.[461] Furthermore, a notable portion of modern literature conforms to the secularized perspective of the natural sciences and lacks concern for the coherence of reality.[462] In modern times, scientism has created a dominant mindset that upholds the worldview of *takthir*. Consequently, even if Muslim rejects certain scientific theories, their awareness cannot escape the influence of scientific thinking. Regardless of their chosen career, they are still likely to adopt the prevailing thought patterns of their profession, which are influenced by scientism.[463]

As discussed in previous chapters, the basis for meaning and ethics in Islam is the concept of *'tawhid'*. According to Nasr, the lack of "principal knowledge" is responsible for the prevalence of *'takthir'* and its consequent loss of direction and purpose. The acceptance of *'takthir'* not only results in external consequences but also represents the most substantial danger to Islamic ethics and values. [464]

Nasr suggests that the potential hazards of the divisive nature of modern science are not only a concern for the West but even more so for Islam, as its fundamental aim is to uphold the "doctrine of unity" (*al-tawhid*) and apply

[460] Grant, "*Technology and Justice*," 35-37.

[461] Nasr, "*A Young Muslim's Guide*," 187.

[462] Nasr, "*Islam and the Plight of Modern Man*," 201-222. In the same vein, Heba Ezzat discusses the effect of secular thought on Muslim social structure.

[463] Chittick, "*Science of the Cosmos*," 11.

[464] Nasr, "*Need for a Sacred Science*," 81.

it to every aspect of life. Therefore, for any Muslim seeking to safeguard genuine Islamic values and the environment, it is essential to be aware of the constraints of contemporary science.[465]

Chapter Summary

Seyyed Hossein Nasr's insights into the ecological crisis reveal that the current scientific worldview, characterized by scientism and scientific progressivism, has contributed to environmental degradation. His arguments show that the shift away from traditional perceptions of nature, which incorporated references to God, to a modern scientific perspective that denies any indication of God's role or existence has led to the emergence of scientism and scientific progressivism. Nasr's proposals for revitalizing religious worldviews to address the environmental crisis involve connecting with Islamic traditions and its rich collection of traditional intelligence on the environment. Through this, Muslims can be motivated to adopt more environmentally friendly behaviors and lifestyles. Unless the issues with scientism are acknowledged, it will likely cause Muslims to adopt a more materialistic lifestyle, which will further exacerbate the environmental crisis. Nasr's arguments highlight the limitations of modern science and emphasize the need to incorporate foundational principles of the unity and hierarchy of reality and the ultimate meaningfulness of the cosmos to develop a long-term solution to the ecological crisis.

The third chapter delves deeper into Nasr's perspective by exploring his conception of Islamic Sharia, metaphysics and its relevance to ecological ethics. It examines the three core concepts of the perennial philosophy in Islamic metaphysics and how they inform Islamic ecological ethics. Furthermore, the chapter explains how contemporary scientific worldviews can undermine metaphysical and religious doctrines that underpin Islamic ethics and how Nasr proposes to address this challenge.

465 Seyyed Hossein Nasr, "Pertinence of Studying Islamic Philosophy Today," in "*Islamic Life and Thought*" (Chicago: ABC International Group, 2001), 150.

Chapter 3

The Islamic Foundation of Nasr's Thoughts

Nasr's perspective is rooted in his understanding of Islamic metaphysics, which he sees as providing a framework for a sustainable relationship between humans and the natural world. This chapter will explore Nasr's conception of Islamic metaphysics, Sharia and their relevance to ecological ethics. This chapter will examine the three core concepts of perennial philosophy – the unity of reality, the hierarchy of reality, and the ultimate meaningfulness or purposefulness of the universe – in Islamic metaphysics, and how they inform Islamic ecological ethics. This chapter will also discuss the ways in which contemporary scientific worldviews can undermine the metaphysical and religious doctrines that underpin Islamic ethics, and how Nasr proposes to address this challenge. Moreover, this chapter will explore Nasr's approach to promoting Islamic environmental ethics, highlighting his emphasis on reviving and adhering to these ethics for environmental preservation. The chapter examines how Islamic theological teachings, or *ma'rifa,* form the basis of Islamic ethics, and how Sufism plays a crucial role in supporting and validating the metaphysical and religious understanding of nature.

Overall, this chapter aims to present a comprehensive overview of Nasr's conventional Islamic response to the ecological problem, rooted in Islamic metaphysics and ethics. We hope that this analysis will shed light on the potential for Islamic thought to contribute to the development of sustainable solutions to environmental problems.

Perennial Philosophy

The concept that all faiths share a similar perennial philosophy—a timeless truth—is not exclusive to Traditionalists. Agostino Steucho, a Vatican librarian in the sixteenth century, is really attributed with coining the phrase "perennial philosophy" when he wrote *De Perenni Philosophia* in 1540. However, the idea dates back much further, to the Latin and Greek Church Fathers as well as Philo of Alexandria's philosophic theology. The notion of a singular ancient philosophy and theology originating from paradise, as proposed by Wilhelm Smidt-Biggemann, has its origins in late antiquity, particularly among the Christian church fathers. It was later embraced by Giovanni Picodela Mirandola, who called it *philosophia prisca,* and subsequently by Nicholas of Cusa, Marsilo Ficino, Giovanni Pico dela Mirandola, and Agostino Steuco during the Renaissance. The Renaissance scholars had to rediscover Hermeticism and Neo-Platonism to fully embrace the concept. In the seventeenth century, thinkers of the Enlightenment, such as Leibniz, Lessing, and Voltaire, also adopted the idea of perennial philosophy. This philosophy often suggested that knowledge had been transmitted through a lineage of great teachers, beginning with Zoroaster, Hermes, Orpheus, Plato, Jesus, and so on.[466] Nasr does not support the humanist interpretation of Renaissance philosophers such as Ficino and Pico. He believes that the philosophers played a crucial role in formulating a distinctive notion of freedom, referred to as "man's inherent freedom from all restrictions," which he contends is the fundamental cause of the contemporary issue. Instead, he endorses the Neo-Platonism and Western Hermeticism perspectives, which he regards as genuine Western customs.[467] Nasr strongly disapproves of the "occult" groups that surfaced during the 19th and 20th centuries, such as the Order of the Golden Dawn, Gurdjieff, and Blavatsky's Theosophy. He also scrutinizes modern spiritual or "esoteric" movements that advocate for the unity of all religions. The Traditionalists widely view Aldous Huxley's popular book Perennial Philosophy as flawed, but it is where most people today first encounter the term "Perennial Philosophy."[468]

A lot of misinterpretation has been caused by the idea that all philosophers and movements that support a "perennial philosophy" can be grouped together under this flag: traditionalists like Coomeraswany and Schuon are herded together with theosophists, anthroposophists, Gurdjieffians, neo-Hindu universalists, neo-Deists, pseudo-mystical romantics, syncretists, and the occult.[469] These movements and ideas are referred to

---

466 Schmidt-Biggeman, W., *Philosophia Perennis: Historical Outlines of Western Spirituality in Ancient, Medieval and Early Modern Thought,* International Archives of the History of Ideas, Dordrecht: Springer, 2004.

467 Nasr, S.H., *Religion and the Order of Nature,* 1996, p131, p173

468 Huxley, A., *The Perennial Philosophy,* Harper & Row, New York: 1970.

469 Oldmeadow, K., *Traditionalism: Religion in the Light of the Perennial Philosophy,* Shrilanka Institute of Traditional Studies, 2000. p158-159

as "pseudo- or counter-traditions." The significance of religion's role and purpose distinguishes traditionalists from progressives. Religions that are "genuine" in the eyes of Traditionalists are those that are founded on revelation. These are "self-sufficient and completely adequate in providing all things necessary to the civilizations in which they have emerged."[470] More crucially, there can be no compromising when it comes to the exterior, or "exoteric," manifestations of revealed religious practice. Orthodoxy, or following the rules of the faith, is a must. A Christian traditionalist must be orthodox, while a Muslim traditionalist must follow the *sharia'ite* laws, which include the five everyday prayers, the Ramadan fast, the Mecca pilgrimage, giving alms (zakat), and other social laws.

The relationship between a person and God is at the foundation of all religions, according to Nasr. Only one part of this relationship is highlighted by each religion, yet regardless of the restrictions imposed by its many forms, each one's principles ultimately include the truth itself. Because of this, fully embracing one religion entails fully embracing all religions. Trying to make a syncretism of various religions universal while actually destroying their revealed forms, which are the sole means by which people can connect with God, is the most futile and harmful action one can undertake.[471]

Although Traditionalists are often associated with the "perennial philosophy" due to its centrality to their beliefs, it is crucial to recognize that their ideology differs from other "perennial philosophies" that employ a comparable term. As Oldmeadow succinctly puts it, the issue of religious forms is the crucial factor in distinguishing Traditionalists from other purportedly perennial philosophers, whose responses could be unfavorable upon hearing words like dogmas or orthodoxy.[472]

The Islamic Tradition and the Perennial Principle

Islam's fundamental principles of *tawhid* (Unity of God), the hierarchical organization of existence, and the significance or intentionality of the

[470] Oldmeadow, K., *Traditionalism: Religion in the Light of the Perennial Philosophy,* 2000. p.144.

[471] Nasr, S.H., "*Islam and the Realities of Islam,*" Allan &Unwin, 1975, p16.

[472] Oldmeadow, K., "*Traditionalism: Religion in the Light of the Perennial Philosophy*", 2000. p.145

universe are crucial and consistent with the core principles of the perennial philosophy. Although these ideas can be perceived in their literal or conventional sense, Islamic metaphysics illuminates their similarity to perennial principles and resolves issues that cannot be resolved by the external aspect alone.

Tawhid (Unity of God)

Islam's fundamental premise is Tawhid, which upholds the concept of reality's oneness. Tawhid signifies God's absolute oneness and unity, as He is eternally One (Al-Ahad) and the Sovereign (Al-Malik) over all things. No one can be compared to Him. While the Quran emphasizes the transcendent aspect of God, it also suggests His proximity (Al-Qarib), love (Al-Wadud), and natural events as evidence of His existence. The verse "Wherever you turn there is God's countenance" implies an eternal relationship between God and His creation. Essentially, Tawhid creates an inherent ontological connection between the existing order and God that cannot be comprehended solely by considering God in His transcendent state.[473] [474]

Hierarchy of Reality

Although Truth is the foundation of all creation,[475] in the cosmic demonstration, this hierarchy is founded on distinctions between the basic created components that were employed to create the various types of creatures. As a result, the created order described in the Quran comprises, among other things, entities formed of earth, fire, and light, including humans, jinns, and angels, with the angelic level being above that of mankind and jinns, respectively.[476] Even among the angels, there is a hierarchy of angels through which heavenly authority passes, with those who are nearer to divine actuality being given more authority.

Thus, the observable universe is depending on the unobserved universe

[473] Qur'an 112:4.

[474] Qur'an 2:115.

[475] Qur'an 14: 19

[476] Qur'an 15:27; 32:7.

and the Qur'an talks of both a hierarchy of reality and a mostly hidden hierarchy.[477] There is a hierarchy of knowing capacity that corresponds to the hierarchy of reality.[478] The Qur'an makes references to different capabilities for interpreting God's messages depending on a person's "understanding" or degree of intelligence.[479] But if our vision may vary depending on our level of "understanding" or wisdom, then it stands to reason that what we see with our senses and common sense is merely the beginning of grasping its greater reality.

Ultimate Meaning or Purpose

God's creations are never made in vain. Every creature and every incident reflects the wisdom, elegance, charity, and majesty of its Creator. All of these creatures are indications that reflect the wisdom, beauty, generosity, and majesty of their Creator. The relationship between a creature's purpose and its function as a sign or representation of God's essence can be comprehended through the metaphysical analysis of the principles of *tawhid* and the hierarchical arrangement of existence.[480] [481] [482]

An Exploration of the Metaphysical Concepts of Tawhid and the Hierarchical Nature of Reality

The issues addressed by the Qur'an demand more thought and an appreciation of God's immanence in the context of His supremacy. The ultimate supremacy of God in relation to His creation appears to be simply one side of the narrative in the Qur'anic allusions to the *tawhid* concepts and hierarchy of creation. Nasr and other Islamic environmentalists frequently

---

477 Qur'an 2:3; 3:44; 50:33.

478 "Are they equal – those who know and those who do not?" (Qur'an 39:9).

479 "We detail our signs for people who have understanding." (Qur'an 6:98). "In the creation of the heavens and the earth…are signs for people who are wise."(Qur'an 2:164).

480 "Not without purpose did We create heaven and earth and all between! " (Qur'an 38:27).

481 "Verily, We have created all things in proportion and measure" (Qur'an 54:49).

482 "Among His Signs is the creation of the heavens and the earth, and the living creatures that He has scattered through them." (Qur'an 42:29).

draw attention to the fact that both natural phenomena and Quranic verses use the term aya (sign or symbol of God), which refers to a link between God and the cosmos.[483] This demonstrates unambiguously that environment is a way to learn about God like the Quran is. It is possible to define Islamic metaphysics (*ma'rifah*) as the comprehensive examination of the understanding that everything around us, not just the Qur'an, is a tool for knowing God. Additionally, Abu Bakr al-Kalabadhi (d. 994), a distinguished Sufi, referenced Junayd al-Baghdadi (d. 910), one of the foremost Sufi scholars, to explain that metaphysics (*ma'rifah*) comprises the immediate self-manifestation of God in His devotees or the genuine comprehension of God's indications (ayat Allah) "in the horizons and within themselves" (Qur'an 41:53).[484] Abd Allah Ansari (1006-88), another prominent Sufi, emphasizes this idea by stating that metaphysics (*ma'rifah*) involves comprehending the essence of a thing in its true form. This implies that from a metaphysical viewpoint, comprehending the "essence of a thing as it is" is equal to understanding God's self-disclosure. Therefore, Islamic metaphysics seeks to unveil the truths of God's messages that shape the universe. As a result, a comprehensive grasp of the cosmos through all facets of metaphysics is imperative, along with a thorough explanation of the principles of unity and hierarchy in creation. As we will see, the concept of purposefulness is intimately connected to these two concepts.[485]

While the concept of tawhid (oneness of God) and the hierarchy of existence were previously alluded to by Sufis, the most thorough theoretical structure for these two metaphysical concepts in Islamic tradition can be found in the Sufi denomination and its affiliated school, which was established by Muhyi al-Din Ibn 'Arabi (1165-1240). Mulla Sadra (1571-1640) was also a part of this school. Nasr asserts that Ibn 'Arabi elucidated the most profound doctrine imaginable

---

483 Seyyed Hossein Nasr, "Reading the Cosmic Qur'an," lecture delivered to the Muslim Students' Association (MSA) at the George Mason University (USA), 27 March 2008.

484 Abu Bakr al-Kalabadhi, "The Exploration of Sufi Teachings (*Kitab at-ta'arruf*)," trans. John Renard in "*Knowledge of God in Classical Sufism: Foundations of Islamic Mystical Theology,* The Classics of Western Spirituality Series" (New York: Paulist Press, 2004), 101.

485 'Abd Allah Ansari, "The Hundred Fields: Resting Places of the Wayfarers," trans. John Renard in "*Knowledge of God in Classical Sufism: Foundations of Islamic Mystical Theology,* The Classics of Western Spirituality Series" (New York: Paulist Press, 2004), 296.

of Being and its manifestations. [486] This immediately connects to the hierarchy of realities and the tawhid. In *The Garden of Truth*, Nasr's previously released book on Sufism, it is clear that Nasr bases much of his own interpretation of Sufism on the ideas advanced by Ibn 'Arabi.[487] According to William Chittick, the primary expert on Sufism and Ibn 'Arabi in the Western world, nearly all subsequent intellectual manifestations of Sufism can be traced back directly or indirectly to Ibn 'Arabi's writings or those of his followers. This underscores the pivotal position that Ibn 'Arabi occupies within Islamic tradition.[488]

Nasr acknowledges Ibn 'Arabi as a prominent proponent of the perennial philosophy, which is highly pertinent to our current subject matter. In order to comprehend how the principles of the perennial philosophy are not only rooted in Islamic metaphysics but also form the foundation of Islamic metaphysics regarding the environment, it is essential to elaborate on the fundamental tenets of Ibn 'Arabi's metaphysics, particularly regarding the interconnection among human beings, God, and the universe, which comprise the entirety of existence.[489]

Islamic Metaphysics and Cosmology

To present Ibn 'Arabi's metaphysical interpretation of tawhid, we will delve into the evolution of three interrelated concepts that have been present in Sufism since his era:

1. The concept of the unchanging prototype (*al-a'yan al-thabita*)

2. The manifestation or self-revelation of God (*tajalli*)

3. The doctrine of the Unity of Being (*wahdat al-wujud*)

---

486 Seyyed Hossein Nasr, *"Islamic Philosophy from its Origin to the Present: Philosophy in the Land of Prophecy"* (Albany: State university of New York Press, 2006), 86.

487 Seyyed Hossein Nasr, *"The Garden of Truth: The Vision and Promise of Sufism, Islam's Mystical Tradition"* (New York: HarperCollins Publishers, 2007), 30-58.

488 William Chittick, "Ibn 'Arabi and His School," "in *Islamic Spirituality: Manifestations,"* ed. Seyyed Hossein Nasr (New York: The Crossroad Publishing Company, 1997), *68*.

489 Seyyed Hossein Nasr, "Reply to Huston Smith", in *"The Philosophy of Seyyed Hossein Nasr"*, The Library of Living Philosophers, Volume XXVII, eds. Lewis Edwin Hahn, Randall E. Auxier, Lucian W. Stone, Jr. (Chicago: Open Court Publishing Company, 2001), 160.

The Concept of the Unchanging Prototype (Al-A'Yan Al-Thabita) and Manifestation of God (Tajalli)

The notion of immutable patterns in Islamic metaphysics derives from the Quranic affirmation that Allah has knowledge of all things prior to their inception. When Allah intends for something to exist, He simply speaks the word "Be," and it comes into being.[490] This "thing" exists as an immutable archetype in the divine realm prior to its manifestation in the physical world. The hadith *qudsi* portrays the purpose of creation as the desire of Allah to be known, saying, "I was a hidden treasure, so I created creatures in order to be known." [491] [492] Ibn 'Arabi interprets the divine command "Be!" as a means for Allah to reveal Himself through His creation, as the created object becomes a "locus of manifestation for the Real." Thus, when Allah says "Be," it is and becomes a way for Him to make Himself known through His creatures. [493]

Put differently, the immutable archetype serves as a medium for God to breathe life into it and bring it into existence. According to Ibn 'Arabi, the "locus of manifestation of the Real" reveals the specific qualities of existing things, which are associated with the heavenly names mentioned in the Qur'an. As for God's essence (dhat), it is beyond human comprehension. He is defined by His qualities, which correlate to His name and constitute His first boundaries. As a result of His yearning to be known, His identities are His first Self-disclosures. Ibn 'Arabi believed that the Qur'an serves as a means for God to reveal Himself (*tajalli*) through language, as He discloses His qualities most directly through His names. While God's nature remains incomprehensible, just as light is invisible but takes on various colors, His characteristics can be understood through His Self-disclosure.

The heavenly terms serve as the formed order's universal prototypes. All

---

[490] Qur'an 16:40. See "*The Message of the Quran*, trans. Muhammad Asad."

[491] "A *hadith* is a saying attributed to the Prophet Muhammad. But a *hadith qudsi* is a saying attributed to God but uttered by the Prophet Muhammad and is not part of the Qur'an."

[492] William Chittick, "*Imaginal Worlds: Ibn al-Arabi and the Problem of Religious Diversity*", (Albany: State University of New York Press, 1994), 29.

[493] Ibn 'Arabi, "al-*Futuhat al-Makkiyya*," vol. II 484.23 translated and quoted in Chittick, "*SPK*, 90. Henceforth *al-Futuhat al-Makkiyya*" authored by Ibn 'Arabi will be referred to as just *Futuhat*.

beings contain the Being of God, yet either they do not all show the identical characteristics of God, or those characteristics do not shine as vividly in each thing. When an unchangeable archetype is granted existence, it authorizes the manifestation of a certain confluence of heavenly names and chooses the intensity of those manifestations. Each archetype that God existential displays a different blend of His qualities. All existing things are, therefore "Self-disclosures" of God (*tajalli*) in Ibn 'Arabi's phrase, according to their unchangeable prototypes in the understanding of God.

One may argue that one of God's Being's unlimited possibilities is an eternal archetype. In other ways, His awareness of a certain potential for His Being is an eternal archetype. This prototype may or may not be actualized by Him, but as a potential aspect of His Being, it must endure and be eternal in His memory forever. Because of this, an entity's permanent archetype does not cease even after it dies on Earth. As archetypes are not real in themselves, we cannot directly perceive the unchanging archetypes of existing entities. Instead, what we sense in the cosmos are the divine characteristics themselves, which become revealed in the created world. According to Ibn 'Arabi, the Qur'an affirms that "We created not the heavens and the earth, and what is between them, save through the Real" (Qur'an 15:85), which refers to Pure Being. Hence, all characteristics associated with Pure Being are attributed to the actualities of entities, while the observed effects are attributed to the heavenly names.[494]

According to this view, the universe represents a manifestation of God's Self-disclosure, making it a theophany. It can be seen as the Cosmic Qur'an (*al-Qur'an al-takwini*), which serves as a counterpart to the Written or Oral Qur'an.[495] But it's important not to mistake this conception of the world as a theophany for its deification. Ibn 'Arabi states that just when something comes into being through the Creative word "Be," it doesn't follow that it acquires its own "being" or that God is no longer transcendent: It is important to note that the object itself does not "acquire existence," but rather becomes a locus of manifestation for God. The object retains its own identity as a separate entity from God, who remains distinct as the divine essence. In other words, God is God, and the created things are distinct

[494] *Futuhat* II 216.7 in *SPK*, 95-96. Also, see *Imaginal Worlds*, 18.

[495] Nasr, *Need for a Sacred Science*, 130-31.

from God as their own individual entities.[496] Every existing creature is similar to a mirror of emptiness that reflects certain attributes of God.[497] To put it another way, a thing's ultimate significance is the characteristics of God shown through it, and its main objective is to present God in that form.

Jalal al-Din Rumi (1207-73) succinctly described the connection between the attributes of the Creator and the created with the following phrases:

> "Think of the creatures as pure and clear water,
> In which the qualities of the Almighty shine.
> Their knowledge, justice, and kindness,
> Are all like stars of heaven reflected in flowing water."[498]

The Doctrine of the Unity of Being (Wahdat Al-Wujud)

While Ibn 'Arabi did not explicitly use the term "*wahdat al-wujud,*" his ideas regarding the nature of God and the universe provide insight into the concept. *Wahdat al-wujud,* or the unity of being, is rooted in the belief that the universe is comprised of immutable archetypes that do not exist independently but were created by a single God in accordance with tawhid, or the belief in the oneness of God.[499] Therefore, the *wahdat al-wujud* theory merely asserts that Being is One although existents are multiple, not that multiplicity is erroneous. Each existent's existence is neither distinct from nor independent of God's Being. All things contain the fullness of the God Being, yet they can only manifest the divine traits dictated by their unchanging archetypes. *Wahdat al-Wujud* asserts the presence of the Being of God alone without dismissing the plurality of existents since the unchangeable archetypes are inherently nonexistent. Thus, Nasr occasionally refers to *wahdat al-wujud* as the "transcendent unity of being."[500] While

---

[496] *Futuhat,* II 484.23 in Chittick, *SPK,* 90.

[497] *Futuhat* III 46.27, 47.25 in Chittick, *SPK,* 205.

[498] Jalal al-Din Rumi, "*Mathnawi* vol. VI 3172-73", translated and quoted in William Chittick, "*The Sufi Path of Love: The Spiritual Teachings of Rumi*" (Albany: State University of New York Press, 1983), 43.

[499] The term "*wahdat al-wujud*" was coined by Sa'id al-Din Farghani (d. 1300), one of Ibn 'Arabi's followers.

[500] Seyyed Hossein Nasr, "*Religion and the Order of Nature*" (New York and Oxford: Oxford University Press, 1996), 283.

Ibn 'Arabi provided an intricate metaphysics of the holy names and characteristics, many Sufi poets also alluded to the existence of *wahdat al-wujud*. Rumi expresses the notion that our existences and ourselves are non-existent, and that it is only the Absolute that appears in the form of mortality.[501] In a similar vein, Ibn al-Farid, a renowned Arab Sufi poet, emphasizes the reality of the one God beyond all veils and actions, stating that everything we see is the act of the One. When the screen is lifted, the forms disappear, leaving only Him.[502]

These ideas are central to the concept of *wahdat al-wujud*, which forms a cornerstone of Nasr's philosophy on the universe as a theophany. This principle aligns with the initial principle of the perennial philosophy, which recognizes a singular Ultimate Principle or Reality as the source from which the plurality of existents in the universe arises.[503]

The Concept of Humans as Both the Image of God and the Microcosm

Exploring Ibn 'Arabi's teachings on the unchanging archetypes (*al-a'yan al-thabita*) and *wahdat al-wujud* (Unity of Being) can provide insight into his perspective on the human being, their relationship with the universe, and their ultimate purpose. The renowned hadith of the Prophet states that God created Adam (the human being) in His image, thus distinguishing them from all other entities. Ibn 'Arabi further elaborates on this by interpreting a hadith *qudsi* and Qur'anic verses, explaining that it is no surprise that humans alone are the focal point of demonstration for all of the heavenly names:

> According to a Hadith Qudsi, God affirms that neither His earth nor His heaven can contain Him, but the heart of His faithful servant can. This implies that all of God's names are only revealed

---

[501] Jalal al-Din Rumi, "*Mathnawi*," quoted in Seyyed Hossein Nasr: "*Islamic Art and Spirituality*" (New York: State University of New York Press, 1987), 138.

[502] Ibn al-Farid, quoted in Seyyed Hossein Nasr, "*Sufi Essays*, 3rd edition" (Chicago: ABC International Group, 1999), 78.

[503] Seyyed Hossein Nasr, "*Knowledge and the Sacred*" (Albany: State University of New York, 1989), 189-214.

> through the human form. As stated in the Quran (2:31), Adam was instructed by God on all the names, including the divine names that are the basis for creating everything in existence.[504]

Ibn 'Arabi provided theoretical justifications for what generations of Sufis had learned through intuition and experience before him. Mansur al-Hallaj famously proclaimed (d. 922), "I am He whom I love, and He whom I love is I." [505]

Farid al-Din 'Attar's Conference of the Birds employs the symbolism of a flock of birds to describe the mystic journey, culminating in the realization of the thirty birds who make it to the heavenly appearance. These birds, known as the *simurgh* in Persian, discover that they themselves are the fabled bird that embodies divinity for 'Attar. Through this realization, they also come to understand that they are merely reflections of the Being they pursued and worshipped, resulting in the disappearance of otherness and duality. The consumption of the view of the existence of the *Simurgh* leads to the realization that one cannot discern whether they are the *Simurgh* or the *Simurgh* is them, as they have merged and duality has vanished.[506]

According to Ibn 'Arabi, the universe is a manifestation of God's names, and the human being is the microcosm of the universe if he is the site of manifestation for all these names. The Qur'an explicitly states the connection between the human being and the universe as God says, "We shall show them our signs upon the horizons and in themselves until it is clear to them that it is the Truth" (41:53). The linkage indicates that humans are a miniature representation of the universe, encompassing the same symbols that exist in the universe.[507] [508]

---

504 *Futuhat* I 216.9 in "*SPK*, 276."

505 Mansur al-Hallaj quoted in "Annemarie Schimmel, *As Through a Veil: Mystical Poetry in Islam*" (Oxford, UK: Oneworld, 2001), 32.

506 Farid al-Din "Attar, *Conference of the Birds*," quoted in S.H. Nasr: "*Islamic Art and Spirituality*", 110.

507 Abu Hamid al-Ghazzali, "*The Niche of Lights*, trans. David Buchman" (Provo, Utah: Brigham Young University, 1998), 31.

508 Ibn 'Arabi , "*Futuhat* vol. II 150.33" quoted in William Chittick in "*Self–Disclosure of God: Principles of Ibn al-Arabi's Cosmology*" (Albany: State University of New York, 1998), 9. (Henceforth referred to as SDG). Also, see Qur'an 51: 20-1.

The universe would be incomplete without humans because, in Ibn 'Arabi's metaphysics, no non-human entity represents the entire spectrum of God's names. Unlike in the human entity, where all the heavenly names appear collectively. Therefore, only the existence of humans in the universe qualifies it as a whole theophany since they are the only creatures in the universe that fully represent all of God's attributes. Ibn 'Arabi concluded that the human person is comparable to the essence of the universe:

> Adam is the all-encompassing book, and the entire cosmos is a manifestation of his differentiation. This relationship is similar to that of the spirit to the body. When a human being is present within the cosmos, it becomes a "great human being" by virtue of the synthesis of all its components.[509]

As a result, not only do the universe and the human represent each other as images of God, but also that the universe and the human being are intrinsically connected. Without regard to the other, the other cannot be understood. Humans need to understand God, whose qualities are represented in the universe, in order to understand themselves. We can precisely identify the signs of God in the universe because of the divine identities revealed by the revelation. However, a human could never understand the meaning of the names without referencing cosmic things and happenings. Therefore, in order to understand ourselves or God, we must consider the universe as containing not just immaterial facts, but also signs of God (*ayat Allah*). God made humans so that He can be recognized as God through them, and the universe is crucial in assisting humans in achieving this goal. And a Muslim can also achieve his ultimate goal of knowing God and His Unity by doing so, by knowing God in this way everywhere one turns (*tawhid*).

The Principle of Hierarchy of Reality as Explained by Ibn 'Arabi

Because the unchangeable archetypes permit the emergence of heavenly characteristics in a variety of combinations and intensities, not all things or signs of God reveal God's attributes in the same way. According to Ibn

509 "*Futuhat* vol. II 67.28," quoted in William Chittick, "*Imaginal Worlds*, 34."

'Arabi's metaphysical theory, the hierarchy of existence in the universe is based on the quantity and potency of manifestations of divine qualities.[510]

As we've seen, every quality of God is reflected in the human being. As a result, there is a parallel hierarchy of reality in the material realm, with humanity at the top and the realms of animals, plants, and minerals below it. The *barzakh* (55:19) is a phrase used in the Qur'an to describe a planet of transitional realities that exist between the observable material sphere and the unobserved realm of God and His angels and has characteristics of both the metaphysical above and the physical realm beneath.[511] This intermediary world is where traditional narratives of the soul's postmortem existence up until its resurrection on the Day of Judgment take place. Despite being immaterial, the *jinns* and other similar entities in this universe have shapes and sensory properties. Ibn 'Arabi uses the word "imagination" to describe this transitional world's actuality.

According to Chittick, Ibn 'Arabi uses the word "imagination" in two overlapping contexts. It can be seen as either the intermediary reality we already mentioned, or In its strictest interpretation, it refers to a distinct ability of the psyche that unites physical objects that possess shapes and forms with shapeless and formless consciousness.[512] The term "imaginal" is typically used by modern Ibn 'Arabi academics to differentiate between the sensation of unreality connected with the concept of "imagination" in its present English usage and the transitional reality or the capacity to recognize it as such.

According to Ibn 'Arabi's metaphysics, God is considered to be the origin of all phenomena, which is a belief shared by other Islamic scholars. The unalterable archetypes of God serve as the foundation for creation from above. To understand the hierarchy of reality, one must first realize that while God is present equally in every living being, His manifestation is not uniform in all levels of reality. The school of Ibn 'Arabi proposes that the hierarchy of reality is established by the uneven revelation of heavenly names in various planes of the universe, based on unchanging archetypes. Ibn 'Arabi characterizes divine reality as having several degrees of trans-

---

[510] Chittick, "*Imaginal Worlds,*" 21-22.

[511] "*Futuhat* III 42.5" and II 390.4 in *SDG,* 258-59.

[512] Chittick, "*Imaginal Worlds,* 54."

parency, which range from the Absolute Presence of God to the physical realm.[513] Spiritual intensity, which is a measurement of the degree of the manifestation of the divine or Light (*al-Nur*) of God, is a way to define the many expressions of Presence underneath the divine reality.

The unchangeable archetypes and heavenly names are the ones that are most similar to the heavenly essence. An unchanging archetype that God decides to give life to "descends," so to say, from the metaphysical realm above into the imaginative realm, where it initially takes on a "shape" that corresponds to its unchangeable archetype there.[514] This imaginary form "descends" to the material realm and becomes married to matter, which is only potentiality, creating the physical entity that corresponds to the imaginary "shape" in the imaginary realm above.[515] The imaginal dimension can be understood as a collection of divine attributes that correspond to an unchanging archetype, but with reduced intensity. Similar to this, when paired with the matter below, a physical entity's apparent shape is nothing more than the outermost expression of the attributes of the imaginal dimension.[516] As a result, the principle of any particular physical entity's existence in the imaginal realm is its unchangeable archetype, and vice versa, the principle of the physical entity's reality in the physical realm is its imaginal form. The only means for a human to reach entirely spiritual reality is through the imaginal world, which serves as a barrier between the physical and the metaphysical.[517] The hierarchy of human perceiving abilities corresponds to the hierarchy of reality. Ibn 'Arabi believed that the only things that may open the capacities of vision for degrees of reality above those observed by ordinary rationality are spiritual activity and Heavenly grace.[518]

The preceding brief explanation of Ibn 'Arabi's interpretation of the function of the heavenly names leads to the following conclusions. Firstly, there exists a hierarchy of reality among physical creatures, which depends on

---

513 Chittick, "*SPK*, 16."

514 "*Futuhat* III 361.5" in Chittick, "*SPK*,122."

515 "*Futuhat* III 46.27" and 47.25 in Chittick, "*SPK.*, 204."

516 Sadra quoted in Ibrahim Kalin, "*Knowledge in Later Islamic Philosophy*" (Oxford: Oxford University Press, 2010), 104.

517 "*Futuhat* III 47.25" in Chittick, "*SPK*, 205."

518 "*Futuhat* I 271.27" and I 319.27 in Chittick, "*SPK*, 168-69."

the number and level of manifestation of heavenly characteristics. Secondly, each physical creature corresponds to a distinct hierarchy of reality that begins with the unchanging archetype in God's understanding, descends to the imaginal form in the imaginal realm, and finally engages with the matter in the physical world. Moreover, each entity's hierarchy results from the decreasing strength of the heavenly names' manifestation at each successive level, as a consequence of the hierarchy.

We must remember a number of characteristics of reality's hierarchical structure. Every level of reality has an underlying principle in the one above it. The only way for a person to approach the spiritual realm is via the intermediary plane of imaginal reality. The root of the link between various levels of reality is God's Oneness since, most crucially, all levels of reality are finally included inside a singular and infinite Oneness that God is. The need for religion with all of its rituals and signs is one of many things that the metaphysics of reality's hierarchic structure lays the groundwork for understanding, as we shall see in the next chapters.

"SHARIA", and Its Relation to Ecological Ethics

Nasr is concerned that the majority of Islamic religious experts (*Ulama*) nowadays are not overly concerned with the environmental problem and are barely familiar with Islamic principles on the natural environment. He believes there is a pressing need for devout Muslims, in particular the "*ulama*," to realize that Islam entails more than just praying to God and acting morally toward other people; it also involves caring for the natural environment.[519]

As a result, Nasr suggests that both madrasahs and contemporary educational institutions in Muslim nations teach Islamic environmental ethics.[520] After acquiring knowledge of Islamic environmental teachings, Muslim

[519] Seyyed Hossein Nasr, "On the Environmental Crisis," interview by Muzaffar Iqbal, in "*Islam, Science, Muslims, and Technology: Seyyed Hossein Nasr in Conversation with Muzaffar Iqbal*" (Alberta, Canada: Al-Qalam Publishing, 2007), 146-47.

[520] Seyyed Hossein Nasr, "Islam, the Contemporary Islamic World, and the Environmental Crisis," in "*Islam and Ecology: A Bestowed Trust*, eds. Richard Foltz, Frederick Danny and Azizan Baharuddin" (Harvard University Press, 2003), 101-102.

nations should collaborate with the "ulama" who have the people's trust and allow them to lead environmental efforts.[521] Furthermore, Nasr encourages modern academics to use the current concrete laws and regulatory principles in Islamic jurisprudence to expand the application of environmental Shari'a and address contemporary environmental issues whenever necessary. [522]

Nasr suggests that Shari'a-approved laws for environmental protection are more likely to be observed by Muslims than secular laws because they view them as divine decrees rather than mere governmental rules that can be disregarded whenever possible.[523] He welcomes attempts to minimize pollution through civil legislation or contemporary NGOs, as long as those initiatives do not contradict Islamic values like legal maxims.[524] Such strategies that conserve the natural environment may count *'ada* and *'urf*,[525] and be approved by Islamic law authorities. In conclusion, Nasr endorses the resurgence of the conventional Islamic waqf system, which involves religious donations for the benefit of the public and is primarily financed by devout individuals, as a means of promoting environmental projects, particularly the establishment and maintenance of harims and *himas*.[526] [527]

In summary, it is possible to envision a comprehensive system of Islamic ethics that, aside from being more likely to be implemented in countries with strong religious convictions, could surpass secular environmental laws,

---

[521] Nasr, "Islam, the Contemporary Islamic World, and the Environmental Crisis," 102.

[522] Nasr, "*Need for a Sacred Science,* 143."

[523] Nasr, "Islam, the Contemporary Islamic World, and the Environmental Crisis," 99.

[524] Ibid., 102; Also Nasr, "*Need for a Sacred Science,* 143."

[525] *'ada* and *'urf* are "customary practices and definitions" in a Muslim society "which may acquire legal force when they accord with the aims of the *Shar*i'a." Othman Llewellyn, "The Basis for a Discipline of the Islamic Environmental Law," 192.

[526] "A *harim* is an inviolate public space or water resource and a *hima* is a protected area for the non-human species. They are among the most important environmental practices in the *Shari'*a. For an excellent discussion of *harims, himas* and suggestions for their extension under modern circumstances," see Othman Llewellyn, "The Basis for a Discipline of the Islamic Environmental Law,"210-217.

[527] Nasr, "Islam, the Contemporary Islamic World, and the Environmental Crisis,"103. The *waqfs* funded the building of mosques, schools, hospitals, wells, and other charitable projects for the public in the pre-modern Islamic world. Llewellyn, "The Basis for a Discipline of the Islamic Environmental Law," 217-8.

particularly in circumstances where they cannot be monitored or enforced by authorities. Nasr argues that these morally motivated initiatives would ultimately fail unless we are able to restore and uphold the spiritual or religious perspective of nature that the scientific paradigm rejects. We must first comprehend the role played by theological and metaphysical beliefs in the development of Islamic ethics in order to fully appreciate this.

The main Islamic theological principles that characterize human-nature interaction are the notions of bestowed trust *(amana)* and vicegerency *(khilafa)*.[528] [529] Furthermore, the Qur'an implies that devotion to God and respect for His Prophet are the correct attitudes required for humans to gain felicity.[530] In fact, service is the only way for a person to function as a *khalifah* (vicegerent) and treat as *amana* the things that God has given him, including his physique, faculties, intellect, soul, and the surrounding natural environment.

Knowing God's will is a prerequisite for being in devotion to Him. The *Shari'a's* primary principles, the explicit divine commandments, reveal God's will, yet everything God has created has essential qualities that reveal God's will immediately and profoundly. Nasr posits that the term *"haqq"* in the Qur'an has multiple meanings, including truth, reality, right, law, and due, and that it is also associated with God, who is referred to as al-Haqq (The Truth).[531] According to Islamic belief, each being exists because of the truth (*haqq*) and is entitled to its due (*haqq*) based on its nature. This applies not only to humans, but also to trees, animals, rivers, and mountains. When interacting with nature, humans must give each creature what it is due, and each creature has its own corresponding rights. These rights are bestowed by God and cannot be taken away by humans at will.[532]

God's intention for how humans should handle nature is ultimately established by that standard since the Qur'an refers to the actuality of creatures

528 See Qur'an 33:72.

529 See Qur'an 2:30.

530 Ingrid Mattson, *"The Story of the Quran: Its History and Place in Muslim Life"* (Malden, MA: Blackwell Publishing, 2008). The verses 3:132; 4:69 asserts the need to obey God and His Prophet to achieve felicity.

531 Nasr, "Islam, the Contemporary Islamic World, and the Environmental Crisis,"97.

532 Ibid.

as evidence of God (*ayat Allah*) (3:190). The concept is emphasized in the Qur'an through the use of a metaphorical representation of the natural environment, the she-camel, which is mentioned several times. According to Mawil Izzi Dien, this representation should not be destroyed due to its inherent value as a symbol of God: [533]

> As the following verse from the Qur'an indicates, the she-camel should be allowed to graze on God's earth and protected from harm, or else those who harm it will face severe consequences: "This she-camel of Allah is a sign unto you: so leave her to graze in God's earth, and let her come to no harm, or you shall be seized with a grievous punishment." (Qur'an 7:73)[534]

All things in the creation of the heavens and the earth are indeed indications of God's existence. By disregarding these signs, we commit a sin against Him and bring about corruption in both the land and the sea.[535]

Our knowledge of Islamic ethics basically rests on a genuine knowledge of the signs, which involves metaphysics (*ma'rifa*), if the right human behavior toward each object in the natural environment is defined essentially by its essence as a sign of God. Taking into account the notions of the immutable archetype (*al-a'yan al-thabita*), the Self-manifestation of God (*tajalli*), and the Unity of Being (*wahdat al-wujud*), it can be inferred that Islamic metaphysics (*ma'rifa*) represents an elaboration of the Qur'anic declaration that all entities are indications of God. However, these convictions are essentially metaphysical elaborations of the conventional religious principles of tawhid and the cosmic hierarchy.

Moreover, the philosophical teachings of the unchangeable archetypes and of the human as the microcosm of the world that represents all the attributes of God who is One make the concepts of *amana* (bestowed trust) and *khilafa* (vicegerency) understandable (*Al-Ahad*). Abu Hamid al-Ghazzali (d. 1111) posited that the *amana*, an inherent capacity of the human being, enables him to comprehend the archetypes of all things within his heart, which is

[533] Izzi Dien, "*The Environmental Dimension of Islam*", 98.

[534] Also, see Qur'an 26:155-158; 17:59.

[535] Qur'an 30:41. See "*The Holy Qur'an*, trans. Muhammad M. Pickthal" (New York: Tahrike Tarsile Qur'an, (2001). Also, see Qur'an 26:155-158.

considered God's dwelling place.[536] Likewise, Nizam al-Din Nishaburi (d. 1327), in his renowned interpretation of the Quran, asserted that humans become God's *khalifah* (vicegerent) when the light of God illuminates their hearts, allowing them to perceive how all things are enmeshed in God.[537] As a result, the principles of *tawhid,* which encompass the order of the cosmos and the purpose of existence, form the basis for the religious doctrines of *amana* and *khilafa*. The foundation of Islamic environmental ethics is largely based on *tawhid,* while the principles of *amana* and *khilafa* guide Islamic ethical considerations towards the environment. Mohammad Hashim *Kamali,* a prominent authority on modern Islamic law (Shari'a) and jurisprudence *(fiqh),* has pointed out that any discussion on Islamic law and ethics must begin with *tawhid*. This is because God created the universe and all its parts are interrelated.[538] [539]

According to Al-Ghazzali, there are four stages of knowledge of *tawhid*. The first stage involves only the declaration of belief in tawhid by the tongue without truly accepting it in the heart. The second stage is characterized by the imitation of others' beliefs. The third stage involves an intuitive understanding of how everything comes from a single source while still acknowledging the existence of multiplicity. Finally, the fourth stage is the highest level, where one sees only One.[540] The first stage of tawhid knowledge, known as the hypocrite's level, involves only superficial belief. The second level is the most prevalent and is based on the beliefs of others. This level of tawhid involves accepting that all aspects of existence must be interconnected because they are all willed or created by the same God,

---

536 See Abu Hamid al-Ghazzali, "The Elaboration of the Marvels of the Heart (*Kitab Shar_ 'aja'ib al-qalb*)," trans. John Renard in "*Knowledge of God in Classical Sufism: Foundations of Islamic Mystical Theology,*" The Classics of Western Spirituality Series (New York: Paulist Press, 2004), 304.

537 Nizam al-Din Nishaburi quoted in Mahmoud M. Ayoub, "*The Quran and Its Interpreters,*" Vol. I (Albany, NY: State University of New York, 1984), 78.

538 "*Tawhid* is known by knowing the rootedness of things in God, the One; the apparent distinction between a thing's ordinary appearance and its root in God proves the hierarchy of reality; and the ultimate meaningfulness of the cosmos again lies in its rootedness in God who is the ultimate reality."

539 Mohammad Hashim Kamali, *Shari'ah Law: "An Introduction"* (Oxford, UK: Oneworld, 2008), 17.

540 Abu Hamid al-Ghazzali, "*On Trust and the Unity of God,*" trans. Mohammad Nur Abdus Salam(Chicago: Great Books of the Islamic World, 2002), 22.

according to Shari'a. [541] Al-Ghazzali proposes that the third and fourth levels of tawhid knowledge reveal connections between things that are not visible to ordinary reason. These levels of knowledge may provide justification for ethical principles or Shari'a injunctions that may seem arbitrary otherwise.

According to Al-Ghazzali, God's Unity and the Religious Law are interconnected, with the Religious Law being a manifestation of God's Unity. He argues that our actions ultimately determine whether we end up in paradise or hell, which are not physical places but rather states within our own genuine reality.[542] Al-Ghazzali suggests that there is a level of awareness of tawhid, or the oneness of God, that enables us to perceive the interrelatedness of everything in the cosmos, including heaven and hell, and see them as part of a larger Unity.[543] He also points out that the hadith "Verily, God created Adam in His image" implies that human beings contain the entire world, including heaven and hell, within themselves as God's image. However, Al-Ghazzali notes that this reality and the integration among people, the non-human realm, actions, and duties are often concealed from our usual consciousness. He believes that acknowledging the truth of this hadith is necessary to comprehend the concept of Unity. Al-Ghazzali further illustrates how metaphysics can aid in understanding the Shari'a, particularly when explanations are unclear. He emphasizes the spiritual idea of human beings as God's image-bearers, inheriting all of His qualities, even if they are unaware of them. He suggests that human beings have the potential to serve as God's vicegerents or ideal representatives on earth.[544] Ibn 'Arabi makes the case that a person can only most accurately reflect God when he is able to fully embody the majesty and kindness of God within himself, being in perfect balance.[545] [546] Later on, we'll discover that Ibn 'Arabi offers a philosophical justification for adhering to Shari'a, outlining how it aids in attaining the optimal equilibrium of divine traits within individuals.

---

[541] Kamali, "*Shari'ah Law*, 18."

[542] Al-Ghazzali, "*On Trust and the Unity of God*, 28."

[543] Ibid., 25.

[544] Ibid. 25 and 33.

[545] See William Chittick, "*Imaginal Worlds: Ibn al-Arabi and the Problem of Religious Diversity*," (Albany: State University of New York Press, 1994), 56.

[546] Ibn al-Arabi, "al-*Futuhat al-Makkiyya*," II 615.22 in William Chittick, "*The Sufi path of knowledge: Ibn al- Arabi's metaphysics of imagination*" (New York: State University of New York Press, 1989), 370.

According to the metaphysics of the hierarchy of reality, the principles governing entities within a given level are situated at the level above. As a result, everything on the physical plane, including beings and meanings, are mere reflections of their higher reality in the imaginal realm and beyond. Since humans can only attain knowledge of metaphysical realities through the imaginal world, those individuals who are connected to higher levels of knowledge are best equipped to convey messages from these realms effectively. They, in turn, can interpret the symbolic or greater reality of events and beings in the physical realm alone. To realize how the Islamic faith, with all of its ceremonies and signs, has reasoning beyond what common reason can comprehend, in Islamic metaphysics, the Qur'an is considered an imaginal reality and its perfect interpreter is believed to be the Prophet, Muhammad. Therefore, it is essential for us to understand this concept first.

By referring to the Qur'an's "descent" as an imaginal reality, Ibn 'Arabi puts emphasis to the imaginal aspect of the scripture. The Qur'an was originally intended to be God's unadulterated instruction. However, it is said that dream visions always came before the Qur'anic revelation.[547] The physical understanding of the Qur'an can be viewed as a combination of the enduring archetype or "essence" of the Qur'an with its audible qualities.[548] Consequently, only individuals who possess access to dimensions beyond the physical realm would be considered capable of understanding the Qur'an in its appropriate interpretation. This leads to the question, what sets the Prophet apart as the preeminent translator of the Qur'an?

The degree of manifestation of God's attributes as the All-Knowing (al-'Alim) or the All-Aware (*al-'Alim*) God determines the differences between degrees of reality. Ibn 'Arabi and other Sufis held that spiritual practice leads to an increase in human knowledge, consciousness, and perfection.[549] Ibn 'Arabi believed that the ideal human perceives God in the visible realm as He exists in the universe, and also perceives Him in the invisible realm

---

[547] See Chittick, "*Imaginal Worlds*," 74. For our discussion on the imaginal realities, see Chapter 2.2.2.

[548] "*Futuhat* II 375.32" in Chittick, "*Imaginal Worlds*, 75."

[549] "*Futuhat* I 271.27" and I 319.27 in Chittick, "*Sufi path of knowledge*", 168-69. On the views of other Sufis, see Chapter 3.4 – 3.4.1.

as He remains hidden.[550] The Prophet thoroughly understood all aspects of reality as the model of the ideal person in Islam. Only the Prophet, who had a complete understanding of the higher levels of reality, could fully explain the teachings of the Qur'an by providing instructions and arranging the most fitting ceremonies and signs since the Qur'an is an imaginary universe in and of itself.[551] This viewpoint asserts that the Shari'a, based primarily on the Prophet's explanations of the Quran, represents the most optimal pathway to God. However, what role does conform to Shari'a law play in establishing a person's balance of divine virtues? [552] The following is a summary of Ibn 'Arabi's reasoning.

The Shari'a is comprised of instructions and regulations, including the fundamental principles of acknowledging God's unity, performing prayers, fasting, giving alms, and making the pilgrimage. These tenets are clearly intended to guide people towards God's compassion and proximity and represent the essence of surrender to God. According to Ibn 'Arabi, the prophets have chosen actions that foster closeness to God, and even prohibitions against lying, stealing, drinking, backbiting, and other negative behaviors are ultimately intended to serve this same purpose. Consequently, it is apparent that the primary objective of the Shari'a is to instill the benevolent characteristics of the pious Muslim.[553]

As stated in the well-known hadith, " I am more merciful than wrathful.," The qualities of mercy are now superior in rank, closer to, or more basic to God. Given that the *Shari'a* prioritizes the development of mercy-related virtues, its adherence helps to integrate, harmonize, and balance the various divine characteristics present in humans.[554] Thus, adhering to *Shari'a* reveals the sincere *Muslim* traits of kindness and harshness in the right proportion, which is an accomplishment that cannot be determined just by human calculations.

---

[550] "*Futuhat* IV 246.12" in Chittick, "*Imaginal Worlds,*" 155. Also, see Chittick, "*Sufi path of knowledge,* 30."

[551] "*Futuhat* III 398.11" in "*Imaginal Worlds,*" 154. Al-Ghazzali makes a similar argument in his "*Al-Munqidh min al-Dalal.*" Abu Hamid al-Ghazzali, *Al-Ghazali's "Path to Sufism: his Deliverance from Error,"* trans. R.J. Mccarthy, S.J. (Louisville: Fons Vitae, 2000), 65-66.

[552] See "*Futuhat* IV 186.22" in Chittick, "*Imaginal Worlds,* 146."

[553] "*Futuhat*" chapter 66 in Chittick, "*Imaginal Worlds,*" 134.

[554] Chittick, "*Imaginal Worlds,*" 3, 148-49.

According to Ibn 'Arabi, Muslims can only achieve the highest level of human potential by following the example of Prophet Muhammad. He firmly believes that the path to true happiness and success lies solely in adhering to the teachings of revealed religion, and nothing else.[555] The rationale for religion becomes weaker when the underlying metaphysics and enduring principles are disregarded or overlooked. By extension, the same reasoning also holds true for the rituals and symbols used by prophets and revelations to represent other faiths.[556] The plurality of religious pathways established by God is recognized by the Qur'an itself. As a result, the Islamic tradition is entirely compatible with the Traditionalist emphasis that followers of each faith observe its ceremonies and symbols.[557] [558]

Sufism and the Knowledge by the Heart

Sufism's primary goal and perspective is metaphysics, also known as *ma'rifa,* which is the knowledge of God or Supreme Reality. It serves as a guide and intellectual understanding of God for those who follow the Sufi path, and as the ultimate experiential knowledge of God. Metaphysical theories also explain the essence of reality for all beings and act as guides to God, the Supreme Reality. It is theorized that these expressions stem from the realized knowledge of individuals who have completed the Sufi journey (al-Haqq).[559] [560]

The well-known hadith *qudsi* was used by Martin Lings, a prominent Islamic intellectual in the twentieth century in the West, to summarize Sufism's aspirations, practice, and doctrine:

---

555 "*Futuhat,* II 148.12," in Chittick, "*Imaginal Worlds,* 146."

556 Ibn 'Arabi recognized not only the validity but also the necessity of diverse religions. See "*Futuhat*" Chapter 48 in *Imaginal Worlds,* 157-160.

557 See the Qur'an 5:48; 22:67; 5:68.

558 See Reza Shah-Kazemi, "*The Other in the Light of the One: The Universality of the Qur'an and Interfaith Dialogue*" (Cambridge, UK: The Islamic Texts Society, 2006).

559 See Ali ibn, "Uthman Hujwiri" (990-1077), "The Revelation of Realities Veiled," trans. John Renard in "*Knowledge of God in Classical Sufism: Foundations of Islamic Mystical Theology,* The Classics of Western Spirituality Series" (New York: Paulist Press, 2004), 273-274.

560 Seyyed Hossein Nasr, "*The Garden of Truth: The Vision and Promise of Sufism, Islam's Mystical Tradition*" (New York: HarperCollins Publishers, 2007), 210-11.

> There is nothing more pleasing to me, as a way for my servant to come closer to me, than the worship that I have made obligatory upon him. And my servant does not cease to draw near to me with voluntary devotions until I love him. And when I love him, I become his hearing with which he hears, his sight with which he sees, his hand with which he grasps, and his foot with which he walks.[561]

Apart from the obligatory duties of Islam defined by Shari'a, Sufism incorporates diverse methods of connecting with God. These include non-obligatory forms of worship such as extra prayers, Qur'an recitation, dhikr (remembrance or invocation of God by His name), spiritual solitude (*khalwa*), and acts of charity that go beyond the minimum requirement for alms-giving. Throughout these practices, the Prophet Muhammad serves as the primary model for emulation. The ultimate goal of Sufism is to approach God and recognize His presence everywhere, as the most excellent way to worship Him is to adore Him as if one sees Him.[562] Sufism places great importance on delving into the internal or deeper reality of the natural world, as it seeks to explore the existence of God in all aspects of life more than any other facet of Islam. Al-Ghazzali asserts that knowledge of God Most High can be attained The heart plays a crucial role in Sufism, as it is believed that only a pure heart can perceive spiritual realities beyond the limitations of reason. This perception, which is often referred to as a vision of God's omnipresence, can only be achieved if the heart is unobstructed by negative qualities and habits. The ultimate goal of Sufism is therefore to attain heart purity. [563] The focus of Sufism, according to al-Ghazzali, is to eliminate hindrances from the soul and eliminate objectionable habits, with the goal of attaining a heart that is exclusively dedicated to God and continuously involved in His remembrance.[564] The idea of the heart as a perceptive organ is rooted in the Qur'an and the Hadith, which stress the

---

[561] *Sahih Bukhari*, Riqaq, 37, quoted in Lings, "*What is Sufism?*," 74.

[562] See Sachiko Murata and William C. Chittick, "*The Vision of Islam*" (St. Paul, Minnesota: Paragon House, 1994), xxv.

[563] Al-Ghazzali quoted on p. 38. In another instance, in seeking to give this vision the Quranic legitimacy, al- Ghazzali quotes the Qur'anic verse "Does not a person, the centre of whose being God has expanded, have the light of his Lord."(39:22). See al-Ghazzali, "Elaboration of the Marvels of the Heart," 305.

[564] Al-Ghazzali, *Al-Ghazali's "Path to Sufism"*,51.

importance of having a heart capable of recognizing spiritual realities both within ourselves and in the natural world. The Qur'an cautions that only those with understanding may interpret God's signs.[565] It highlights the heart as a bodily organ as well as a tool for comprehending a higher degree of reality that is inaccessible to the senses and common sense. Although wrongdoers may possess hearts, they do not comprehend them, as bodily organs and perceptions are inadequate:

> Their hearts are capable of understanding, yet they do not comprehend; their eyes are capable of seeing, yet they do not perceive; their ears are capable of hearing, yet they do not listen. They are more misguided than cattle, as they are heedless. (Qur'an 7:179)[566]

As per the Qur'an, having faith in God and purifying one's heart from materialistic attachments, doubts about God, and evil tendencies can transform the bodily heart into a tool for comprehending higher truths. A clean heart is rewarded with certainty in understanding.[567] his commentary on the Qur'anic verse "The heart did not lie as to what it saw" (53:11), the well-known Sufi Abu Talib al-Makki (d. 996) explains the Qur'anic credibility of the heart's encounter with certainty. He emphasizes that God establishes the heart as having eyes, and therefore the vision of the heart leads to certitude, which is endowed to those who possess such a heart. In Sufi terminology, the Intelligence that we have described is also referred to as the "eye of the heart" or *'ayan al-qalb*.[568]The superiority of intelligence by the clean heart has been proclaimed by Sufis throughout history. Mahmud Shabistari (d. 1321) is a well-known Sufi poet who holds the same view as Rumi,[569] Ibn 'Arabi, al-Ghazzali, and many other Sufis who came before him: that reason alone cannot disclose the reality of matters. Ratiocination, according to Shabistari, may be summed up as the process by

---

565 Qur'an 2:269; 3:7; 3:190; 6:98; 7:32.

566 Also see Qur'an 22:46.

567 Qur'an 9:108; 26:89; 29:49; 7:33; 3:14; 6:32; 64:11.

568 Abu Talib al-Makki, "The Sustenance of Hearts (*Qut al-qulub*)," trans. John Renard in "*Knowledge of God in Classical Sufism: Foundations of Islamic Mystical Theology,* The Classics of Western Spirituality Series" (New York: Paulist Press, 2004), 198.'

569 Jalal al-Din Rumi, "*Mathnawi*" vol. I, 3484-88, trans. William Chittick, in Chittick, "*Sufi Path of Love: The Spiritual Teachings of Rumi*" (Albany: State University of New York Press, 1983), 38.

which the mind first begins with conceptions, then assesses and classifies other notions, and then uses Aristotelian syllogistic reasoning among the "known" conceptions to derive a new "known" from those conceptions. Therefore, knowledge gained by ratiocination is simply an "imitation" and not actual knowledge because it is dependent on other "knowns" and therefore not actual knowledge. Shabistari, however, speaks of the "work of the Heart," which is the only thing that can go beyond the realm of logic:[570]

> The actions of water and clay do not lead to the knowledge that comes from the heart.[571]

This idea has been expressed by many Sufis who use the term "eye of the heart" to describe the heart's ability to perceive all levels of reality. For example, some have said:

> "With the eye of the heart, I saw my Lord and asked, 'Who are you?' He replied, 'You are me!'"[572] Others have said, "The eye of the heart perceives what it sees through the collyrium of His light and mercy, all the way to the seventh heaven." [573] Still, others have said, "Open the eye of the heart so that you can see the spirit and gain vision of that which is invisible."[574]

Nasr cautioned that although intelligence has been limited to reason in contemporary philosophy, the heart has also come to be linked with simply feelings and sentiments rather than the most profound understanding as suggested by the Sufis.[575] The intellect, as the eye of the heart, differs from conventional, dry reasoning in that it possesses both certainty and love. According to the renowned Sufi Shaykh Ahmad al-'Alawi (1869-1934),

---

[570] Sa'd ud- Din Mahmud Shabistari, "*The Garden of Mystery* (*Gulshan-i raz*)," Trans. Robert Darr. (California: Real Impressions, 1998). Verses 73-79.

[571] Shabistari, "*Garden of Mystery*", verses 428-429.

[572] Mansur al-Hallaj (d. 922) quoted in Annemarie Schimmel, *As Through a Veil: Mystical Poetry in Islam* (Oxford, UK: Oneworld, 2001), 32.

[573] Jalal al-Din Rumi , "*Diwan-i Shams-i Tabrizi* 21214," in Chittick, "*Sufi Path of Love,*" 75.

[574] Hatif Ishfahani (d. 1784) quoted in Seyyed Hossein Nasr, "*Islamic Philosophy from its Origin to the Present: Philosophy in the Land of Prophecy*" (Albany: State University of New York, 2006), 102.

[575] Nasr, "*Islamic Philosophy from its Origin to the Present,*" 102.

the Intellect serves as the connection that binds a person's soul to God through true and affectionate intelligence. [576] This is because the Intellect only perceives God, who is known as the Compassionate (Al-Wadud), in everything it perceives. Our intelligence is like a beautiful, flawless jewel that is intoxicated with the wine of love, making us appear mad to others, yet we are not truly mad. Although we may seem like ordinary people, our spirits shine with a brilliant clarity that transcends even the highest heights. This bond with God is like a glimmering light, yet it is incredibly strong.[577]

In order to open the eye of the heart, what is the most effective way to purify it? While mandatory acts of worship are important, they only take up a small amount of time each day. This is why supererogatory deeds are particularly significant for Sufis who wish to keep their hearts focused throughout the rest of the day.[578] Among these deeds, the most common is dhikr, which involves calling God by one of His many names. The importance of dhikr is highlighted in many verses of the Quran and prophetic traditions.[579] [580] The purpose of spiritual practice is to purify the soul of negative traits like greed, envy, jealousy, and hatred. This is accomplished by cultivating virtues that embody the divine qualities represented by God's sacred Names. According to many Sufis, the process of developing virtues through remembrance of God involves assuming the characteristics of His divine Names, which is known as *"altakhalluq bi'l-asma' al-ilahiyya"*. As Rumi eloquently expressed, invoking God's Name brings about blessings, and the Name itself becomes a manifestation of its qualities.[581] [582] [583]

---

576 Martin Lings, *"A Sufi Saint of the Twentieth Century: His Spiritual Heritage and Legacy"*, 3rd edition (Cambridge, UK: Islamic Text Society, 1993), 214.

577 Shaykh Ahmad al-'Alawi quoted in Lings, *"Sufi Saint of the Twentieth Century,"* 214. (accent ours).

578 Lings, *"What is Sufism?,"* 78.

579 See Annemarie Schimmel, *"Mystical Dimensions of Islam"* (Chapel Hill: The University of North Carolina Press, 1975), 167-178.

580 See *"Essential Sufism,"* eds. James Fadiman and Robert Frager (New York: HarperCollins Publishers, 1997), 102.

581 Titus Burckhardt, *"An Introduction to Sufi Doctrine"*, Trans. D.M. Matheson, (Lahore: Sh. Muhammad Ashraf Publishers, 1996), 115-16.

582 William Chittick, *"Sufism: A Short Introduction"* (Oxford: Oneworld Publications, 2001), 57.

583 Jalal al-Din Rumi, "Diwan-i Shams-i Tabrizi", verse 30701, trans. William C. Chittick, in Chittick, *"The Sufi Path of Love: The Spiritual Teachings of Rumi"* (Albany: State University of New York Press, 1983), 159.

Sincere dhikr, also known as the light that appears in the heart according to al-Ghazzali, is crucial in allowing the Intelligence to sparkle and comprehend spiritual realities, including those of nature, by overcoming personal consciousness conditions that obscure it. These conditions encompass honesty, perseverance, fulfilment, benevolence, compassion, and love.[584] The Sufi Hakim at Tirmidhi (d. 908) observed that dhikr sustains *ma'rifa*, and without much conviction, moral teachings of religion lack a compact foundation, and its qualities cannot flourish as easily. Abu Nasr as-Sarraj, a Sufi who lived until 988 CE, suggested that the Revealed Law encompasses two fundamental notions: transmission (*riwaya*) and comprehension (*diraya*). He argued that the cultivation of internal virtues through the Sufi path is essential for achieving a comprehensive and enlightened understanding of the Law.[585] [586] Sarraj emphasized that the outward cannot exist independently of the inward. Similarly, Nasr drew a connection between the Sufi notion of intellect and the Law's origin, indicating that the absence of the former would certainly influence people's grasp and affection towards the Law.[587] [588]

He believes that a significant portion of prophecy can only be understood through experiential knowledge gained by following the path of Sufism. By practicing Sufism, one can better adhere to Islamic principles in all situations. As per an anonymous author of the famous 10th-century manuscript, "The Manners of Kings," Sufis are the group of individuals most closely associated with the prophetic model in all aspects, including their convictions, intentions, and conduct, both externally and internally, publicly and privately.[589] [590]

---

[584] Abu Hamid al-Ghazzali, *"The Niche of Lights"*, trans. David Buchman (Provo, Utah: Brigham Young University, 1998), 26-27.

[585] See John Renard, Introduction to *"Knowledge of God in Classical Sufism: Foundations of Islamic Mystical Theology"* (New York: Paulist Press, 2004), 25.

[586] Abu Nasr as-Sarraj, "The Book of Flashes (*Kitab al-luma'*)," trans. John Renard in *"Knowledge of God in Classical Sufism: Foundations of Islamic Mystical Theology"*, The Classics of Western Spirituality Series (New York: Paulist Press, 2004), 82.

[587] Ibid., 83.

[588] Seyyed Hossein Nasr, *"Traditional Islam in the Modern World,"* (London and New York: Kegan Paul International, 1994), 224.

[589] Al-Ghazzali, *"Al-Ghazali's Path to Sufism,"* 62.

[590] *"Abd al-muluk fi bayan haqa'iq al-tasawwuf"* "(The Manners of Kings: Explaining the Realities of Sufism)", eds. Bernd Ratke, Beiruter Texte und Studien, 37 (Beirut/Stuttgart: Franz Steiner Verlag, 1991), pp. 5-6, quoted in Carl Ernst, *"The Shambala Guide to Sufism,"* 25.

Sufism provides a counterweight to the contemporary scientific outlook, which attempts to grasp the universe's entirety through its metaphysical convictions and the *tahqiq* approach to substantiating alleged facts. Hence, Nasr contends that Sufism represents the most potent means for Muslims to confront the scientific paradigm responsible for ecological problems. While Nasr acknowledges the significance of religious ethics, he argues that they are insufficient, and it is necessary to revive the religious understanding of the natural order, which entails both knowledge and ethics.[591]

When one's heart is purified through Sufi practices and infused with divine attributes, humans can serve as conduits for heavenly blessings to all of nature. Through the lens of a pure heart, nature is perceived as a reflection of the deeper realities of the human spirit, and a means to experience God's beauty, generosity, love, and knowledge.[592] Such a person is compelled to coexist with nature. On the contrary, the Prophet taught that when an evil person dies, nature is relieved.[593]

Sufis became well-known and spread their teachings mainly through their vast collection of authentic poems that extolled the omnipresence, majesty, compassion, and wisdom of God, as manifested in the natural world. Such poetic verses have traditionally been regarded as the most renowned and vital genre of literature in the Islamic tradition. Nasr contends that the revival of Sufi poetry has the potential to serve as a potent tool in rekindling the Islamic view of nature as a sacred manifestation.[594]

The "forms" or outward manifestations of nature in Sufi poetry act as guides rather than as antagonists to the "meanings" they represent. The obvious form-meaning paradox is comparable to other paradoxes found frequently in Sufi poetry, such as exterior and inner, apparent and concealed, visible and unseen, shell and kernel, and so on. Rumi considers the significance of shapes while keeping in mind the Prophet's instruction:

---

591 Seyyed Hossein Nasr, "*Religion and the Order of Nature*" (New York and Oxford: Oxford University Press, 1996), 273.

592 Seyyed Hossein Nasr, "Contemplation and Nature in the Perspective of Sufism," in "*Islamic Life and Thought*" (Chicago: ABC International Group, Inc., 2001), 203.

593 See "*The Translation of the Meanings of Sahih al-Bukhari*," Vol 8, Book 76, Number 519, trans. Muhammad M Khan (Riyadh: Dar-us-Salaam Publications, 1997), 342.

594 Nasr, "Islam, the Contemporary Islamic World, and the Environmental Crisis," 95, 100.

> one should observe the shape of the heavens and the earth and derive benefit from the Universal Meaning inherent in them.[595]

According to Shaykh Muhammad ibn al-Habib (1876-1972), the natural world is intended to aid in the comprehension of God's attributes and His oneness. To achieve this understanding, individuals must contemplate the beauty of the land and sea, which reveals the beauty of Allah both externally and internally. The most significant proof of Allah's boundless perfection is within oneself and the distant horizon. By meditating on the precise arrangements of physical bodies, such as a string of pearls, the diversity of plants and rugged terrain, and the mysteries of the heavens, including the Throne, Foot-stool, and the spirit sent by command, one can acknowledge the reality of tawhid and discard delusions, doubts, and otherness.[596] Rumi echoes this sentiment by noting that God's signs are manifested in all six directions, and those with illuminated eyes can contemplate the gardens of divine beauty in every animal and plant.[597]

Therefore, wherever one turns, there is the face of God. Finally, nothing in the physical universe lacks the essence of God. Therefore, individuals who have an affection for nature do so because it originates from God, and the natural world is blissful due to His presence.[598]

One of the foremost experts on Sufi poetry in the 20th century, Anne-Marie Schimmel, noted that every member of the creature and botanical kingdoms serve as a symbol of God's majesty in Sufi poetry:

> According to the Qur'an, the early Sufis believed that all things in existence were created to worship God. They had a unique way of listening to nature, including trees, flowers, birds, and fish, as each spoke its own language, known as "lisan ul-hal," or the tongue of its whole being. One notable Sufi, Sana'i, who passed away in 1131,

---

[595] Rumi, *"Fihi ma fihi,"* 39/40, in Chittick, *"Sufi Path of Love"*, 20.

[596] Shaykh Muhammad Ibn al-Habib, *"The Diwan of Shaykh Muhammad Ibn al-Habib"* (Cape Town: Madinah Press, 2001), 69, in Fazlun Khalid, "Applying Islamic Environmental Ethics," in *"Environmentalism in the Muslim World"*, ed. Richard C. Foltz (New York: Nova Science Publishers, 2005), 90.

[597] Rumi, *"Mathnawi,"* VI 3640-42, in Chittick, *"Sufi path of Love,"* 306.

[598] The Sufi poet Sa'di (d.1292) quoted in Nasr, "Islam, the Contemporary Islamic World, and the Environmental Crisis," 95,

> composed the beautiful Litany of the Birds. This work portrays how each bird addresses God in its own language. For example, the stork expresses God's ownership of the kingdom and His glory by constantly uttering "lak lak," while the dove repeatedly calls out "ku ku" to inquire "Where? where? "[599]

Similarly, Schimmel explains how mountain, sea, moon, sun, and each bird, flower, animal, and valuable stone has a specific value in Sufi poetry because of how it worships God or because of the message it conveys from Go:[600]As a Sufi and someone who has been greatly impacted by Sufi poetry, Nasr expresses in moving poems his view of the divine presence in nature and shows those who seek it the continuing power of the Sufi vision of nature:

> The beauty of all creatures on earth reflects your divine essence. Whether it's the lovely countenance of a maiden or the graceful flight of birds, the vast expanse of the blue sky or the turbulent sea, the majestic mane of a lion or the colorful sea urchin, your beauty is evident. I hear your beautiful voice in the enchanting song of the whale and the melodious trill of the nightingale in the garden, praising you in the morning chorus. But above all, I witness your beauty in the pure souls of your true devotees, who bask in your radiance and are devoted to your love. [601]

Chapter Summary

This chapter has highlighted the significant role of Islamic Sharia, metaphysics and ecological ethics in understanding the universe as theophany and the connection between humans and nature. The concept of *Wahdat al-wujud* and the perennial philosophy is the foundational principles of this understanding, which is essential to the Islamic worldview. The chapter also emphasizes the need to revive the traditional Islamic approach to

---

599 Annemarie Schimmel, "*As Through a Veil: Mystical Poetry in Islam*" (Oxford, UK: Oneworld, 2001), 75.

600 Ibid., 76-78. Also, see Schimmel, "*Mystical Dimensions of Islam*," 306-309.

601 Seyyed Hossein Nasr, "*The Pilgrimage of Life and the Wisdom of Rumi*" (Oakton, VA: The Foundation of Traditional Studies, 2007), 23.

nature and to promote adherence to Islamic environmental ethics through the guidance of Sufism. It is essential to recognize that the ecological crisis is not merely a scientific or technological issue but has its roots in the spiritual and ethical crisis of modern society. Therefore, it is necessary to uphold and practice Islamic environmental ethics to address the ecological crisis effectively. Overall, the chapter underscores the critical role of Islamic theology, metaphysics, and spirituality in addressing the ecological crisis and promoting sustainable development.

The next chapter will discuss Seyyed Hossein Nasr's Islamic environmental thoughts, which are rooted in traditionalist Islamic beliefs and influenced by the concept of Traditionalism. It also discusses the relationship between Nasr's ideas and Islamic mysticism or Sufism, as well as the concept of esotericism. The chapter provides an in-depth analysis of Nasr's environmental message and his call for the revival of Islamic environmental ethics within a broader philosophical and religious context.

Chapter 4

Nasr's Viewpoints of Religious Ecological Ethics

This chapter will explore the Islamic environmental thoughts of Seyyed Hossein Nasr, a prominent Islamic scholar who has devoted much of his life to the study and practice of Sufism. However, before delving into his environmental message, we need to understand the broader context of his Islamic beliefs and worldview. This will involve examining Nasr's understanding of religion and Islam, as well as exploring the concept of Traditionalism or Perennial Philosophy, a Western philosophical movement of the 20th century that heavily influenced Nasr's thinking. This chapter will focus on traditionalist Islam, particularly Islamic mysticism or Sufism, which Nasr views as the heart of Islam. To help us better understand Nasr's ideas, we will discuss the concept of esotericism, as defined by Antoine Faivre, an expert in this field. Finally, this chapter examines how all of these concepts relate to Nasr's environmental message and his call for the revival of Islamic environmental ethics. By the end of this chapter, readers will have a better understanding of Nasr's Islamic environmental teachings and the broader philosophical and religious context in which they are situated.

The Basic Characteristics of Nasr's Thought Towards Ecological Ethics

> Now that we have covered the fundamentals of Sufism and traditional Islam, in what way is Nasr's ecological ethics connected to this? What sort of ecological ethics is he considering? Two viewpoints can be identified in accordance with the Traditionalist methodology. The first addresses the external aspects of Islamic practice, while the second addresses its inward aspects.

Nasr's conception of "traditional Islam" serves as the primary foundation for his environmental ethics. 'Powerful and convincing religious teachings about the natural environment and the relationship of humans to it,' as well as 'specific commands for human behavior,' are available to the conventional world of Islam, according to him. Islamic environmental origins and teachings can be found in the Qur'an, *Hadith,* and *Sharia,* as well as other Islamic ethical literature. Nasr highlights the extensive natural history of Islamic science, Sufism, and philosophy, which the major schools of theology have largely overlooked when it comes to developing a theology of nature. Numerous significant teachings may also be found in Islamic literature, architecture, landscaping, and urban planning. These truths require rediscovery, revival, and broad dissemination. Although Nasr acknowledges that much of this work remains incomplete, he does offer some fundamental principles of Islamic environmental ethics that can be originated from the Quran, the Hadith, and the Sharia in his book "Islam and Ecology: A Bestowed Trust" (2003), which is part of the Harvard series.[602]

Quran: In the Qur'an, the universe and humanity are both described as having a significant position. According to Islamic belief, the universe was God's initial primal revelation, and as such, evidence of God's existence can be found throughout the unspoiled natural world. According to traditional Islamic philosophy, the cosmic Qur'an is regarded as a complementary counterpart to the scriptural Qur'an. This suggests that both the verses of the Qur'an and the natural world are seen as manifestations of God's signs (ayat). The Qur'an 41:53 confirms this idea by stating that "We will show them Our signs in the horizons and within themselves until it becomes clear to them that it is the truth."

Nature displays both God's knowledge (*hikma*) and will (*irada*). God did not create nature just for human purposes. It serves as a reflection of God's creative power. The universe's blood vessels and arteries are filled with grace or *baraka*. Nature contains deep depths of importance and meaning, much like the Qur'an. Beyond mere factual events, natural occurrences are essentially metaphors for the mental and emotional experiences of a higher reality that is beyond our perception. They serve not only as a means of accessing spiritual presence but also as a way of understanding and contemplating divine wisdom.[603] The purpose of creating humans was to serve as conduits for the cosmic environment surrounding them. Nature

---

[602] Nasr, S.H., "Islam, the Contemporary Islamic World, and the Environmental Crisis' in: *Islam and Ecology: A Bestowed Trust,"* Harvard University Press, 2003. p.85-105

[603] Ibid., p96

is essential to human life because it provides food, shelter, and most importantly, nourishment for our spirits.

Nasr also highlighted the importance of the Qur'anic concept of *haqq*, which encompasses the ideas of truth, reality, right, law, and due in Arabic. According to this concept, every living being has a right or a due in accordance with its nature. The concept is closely linked to the notion of order discussed earlier in the first chapter, which affirms that each creature holds a distinct position in the hierarchical order of existence that has been established by God. As both servants (*'abd*) of God and His vicegerents (*khalifat Allah*), human beings are responsible for upholding the rights (or dues) of all creatures and treating them with respect.

Hadith: Nasr notes that the Prophet's teachings and acts are abundantly recorded in the literature of Hadith. According to legend, the Prophet promoted planting trees, forbade the removal of vegetation and the contaminating of water supplies, loved animals, created areas that were protected from human interference, opposed wasting and the needless destruction of nature based on avarice or greed, and placed a strong emphasis on cleanliness.

Sharia: There are several prohibitions pertaining to the environment in the Divine Law, also known as the *Sharia*. In his work, Nasr talks about various topics related to the ethical treatment of animals, the management of communal natural resources like water, forests, and grazing lands as public assets, and several economic regulations such as curbing usury, limiting excessive consumerism, and preventing the gathering of wealth. The Sharia presents a combination of specific rules and general principles that can be applied to address environmental crises. Nasr stresses that this is particularly important because Muslims are more likely to accept these directives based on the Sharia than they are to comply with secular environmental laws.

For likely similar reasons to those previously mentioned, Nasr does not clarify any of these injunctions' specifics or their ramifications for the social, political, or economic order. His Islamic environmental teachings have the ability to change lives, which should not be understated. The existing global economic and financial institutions are predominantly focused on capital interest and wealth expansion. As a result, a ban on usury could have profound structural and systemic consequences. These Sharia ecological injunctions are now being worked on by some of his pupils and other academics. Othman Abd-ar-Rahman Llewelyn's 2003 article in Islam and Ecology offers valuable insights into the emerging area of Islamic Shari'ite Environmental Law.[604] To create particular environmental injunctions, he draws on broad Islamic legal ideas like *maqasid* or "ultimate purposes." Definitely worth looking at in depth in this field. As we have already determined, Nasr's self-perception as the revitalizer of traditional Islam and its ecological teachings leads him to place great emphasis on the broader perspective:

> Environmental law alone is insufficient. A profound understanding of nature is essential for the creation of environmental laws. It is noteworthy that contemporary Islamic modernist and Reformist movements, such as Wahhabism, demonstrate little reverence for nature, despite their devout adherence to the five daily prayers and Islam. Conversely, traditional Islam offers a comprehensive perspective on nature. My focus has always been on revitalizing traditional Islam and promoting its teachings regarding nature.[605]

In an interview held in October 2008, Nasr expressed his thoughts on the importance and objectives of divine law, and also provided some specific illustrations:[606]

> Firstly, it is important to acknowledge that the Divine Law is the ultimate law and it becomes apparent in human existence and on earth through the Islamic Sharia. However, due to human imperfections, it is never perfectly implemented. Nonetheless, Divine Law remains an essential element, and significant aspects of individual life, such as marriage, divorce, and economic functions, are still governed by it.

---

604 In *"Islam and Ecology"* (2003) he outlines the basis for a discipline of Islamic Law in Llewelyn, O. Abd-ar Rahman, "The Basis for a Discipline of Islamic Environmental Law" in: Foltz. (eds), *Islam and Ecology,* 2003.

605 "From interview with Seyyed Hossein Nasr on 31 October 2008."

606 Ibid

> One of the essential aspects of Sharia is its prescriptions regarding the treatment of nature. Although not extensively researched by the speaker, some of their students have pursued PhDs on the matter. The speaker, however, has focused more on the theoretical, metaphysical, and cosmological aspects of Sharia.
>
> The Sharia includes strict injunctions regarding the use of water, particularly that which cannot be privatized, such as mountain water. Additionally, there are strict regulations concerning the treatment of animals, cutting of trees, and pollution of water. Furthermore, the Sharia emphasizes the avoidance of excessive materialism, which is at odds with consumerism.
>
> In the United States, consumerism has recently suffered a decline, and some people are attempting to cope with it through cosmetic means. However, the speaker believes that a deeper shift is taking place.

Referring to the regulations that stem from the classical (traditional) Sharia law:

> In Islamic society, some laws are influenced by international and European laws. Although Sharia law is applied to a significant extent in Islamic countries, it is not flawlessly implemented. Despite this, devout Muslims still believe that Sharia law is the code they should abide by in theory, and they do follow it. Therefore, believers can be motivated to comply with various requirements.

Nasr places consistent emphasis on the crucial role of the Ulama, traditional scholars, in addressing the issue of environmentalism:

> If the preacher delivers a message every Friday based on Divine Law and Sharia that one should not harm animals, such as pulling a dog's tail on the street or committing acts of animal cruelty, it can have a positive impact.

There are specific and pressing actions required in the earthly realm as per the principles of Islam and Ecology. These proposals are directed towards the Muslim community and are outlined below:

> The Muslim world should engage in a critical evaluation of the scientistic view of reality, which can be stifling, and demonstrate how it goes against the authentic Islamic and broader religious perspective.
>
> It is essential to develop an Islamic understanding of the natural environment and its relation to humanity in a way that can be easily understood by contemporary Muslims, from those in rural areas to philosophers.
>
> At every educational level, it is essential to introduce environmental courses to equip upcoming leaders of the Islamic society, including students at traditional madrasas and those pursuing religious scholarship (ulama). Governments should motivate religious scholars and media to raise awareness about environmental issues. Non-governmental organizations, both Western and Islamic-compliant, such as waqf funds (religious endowments), should expand throughout the Islamic world. It is vital to promote indigenous or alternative technologies in the Muslim world.

Although these recommendations are centered on the Muslim community, it is unfortunate that Nasr did not provide any specific actions for Muslims living in Western countries or for the Western world as a whole.

> Currently, the responsibility mainly lies with individuals and small groups, which have the potential to expand in the future. It is crucial to acknowledge not only the physical consequences but also the spiritual and religious aspects of the environmental crisis. Moreover, it is essential to revive the genuine Islamic viewpoint with strictness and lucidity, without conceding any compromises. It is crucial to urge those who can awaken to recognize that the contemporary world is on the verge of a catastrophe and needs just one more step to fall into ruin. A higher level of consciousness will result from heightened awareness. The teachings of Islam regarding God, humanity, and nature urge individuals to awaken from the dangerous delusion of scientism and humanity's egocentric control over the environment. By embracing these teachings, Muslims can develop a harmonious relationship with nature, while

> also helping the Western world rediscover its overlooked legacy concerning the role of human beings in God's creation.[607]

Nasr's conclusion is that we should strive for an awakening that results from proactive human effort, rather than waiting for the harsh awakening that comes with ecological disasters.[608]

Nasr has also provided us with the Keys of Traditional Wisdom, Spiritual Instruction, and Ritual as further solutions. The Garden of Truth described the imagined Islamic spiritual Pathway, which we determined to be motivated by a traditionalist understanding of Ibn Arabi. The environmental situation is barely mentioned in The Garden of Truth, and no specific ecological Sufi beliefs, ethics, or practices are explained. This may be seen as a sign that the author does not anticipate his audience to be very interested in the subject. We have discovered that Nasr's explanations of the Islamic concepts in *Religion and the Order of Nature* greatly coincide with the Sufi metaphysics as it is presented here. These closely resemble the broader 'Perennialist' idea we've already covered, but now in Sufi guise. As an illustration of the spiritual elite's potency and function:

> Islamic esoteric teachings include detailed explanations about the spiritual hierarchy that upholds the observable universe and the concept of walaya. In Sufism, walaya is typically translated as 'sanctity' and governs the world invisibly. Without it, the order of nature would deteriorate into chaos, and the world would be in disarray. Practically, this means that believers also subscribe to the belief that God sends saintly men to Earth, who, through their presence and practice of rituals and prayers, maintain the social order and preserve the natural order.[609]

What about actual religious rituals and practices? Do the rituals and prayers have a special ecological aspect? Nasr mainly keeps us in the dark in this section. Maybe the exoteric forms of ritual in traditional Islamic practice are adequate in and of themselves. We recall Nasr's discussion of the role

---

[607] Nasr, S.H., "Islam, the Contemporary Islamic World, and the Environmental Crisis"' in: *"Islam and Ecology: A Bestowed Trust," Harvard University Press,* 2003. p.104

[608] Ibid., p.104

[609] Nasr, S.H., "Religion and the Order of Nature", p.281

that traditional ritual plays in restoring equilibrium and maintaining the established order. But in addition to the customs that God has established, orthodoxy also permits other customs. Numerous schools and masters have created highly unique spiritual practices and methods, such as nature retreats, dancing, reciting the heavenly Names, etc. For example, can spiritual presence in 'virgin nature' play a specific function in spiritual practice, prayers (dhikr), or events? Are followers instructed to 'feel the sacred' in virgin nature? What procedures are employed? Is it necessary to evoke certain symbols or pictures in order to reestablish a healthy connection with the environment? As far as I know, Nasr values the love for untouched nature greatly in his life. However, he doesn't seem to explicitly emphasize this aspect of spirituality as a practice.[610]

Traditional Islam

The concept of "traditional Islam" appears at the beginning of Nasr's 1985 book *Traditional Islam in the Modern World.*[611]

> Traditional Islam, which has been observed for centuries by theologians, jurists, philosophers, scientists, artists, poets, Sufis, and common believers across the Islamic world over the last 14 centuries, remains the dominant form of Islam. This variant of Islam is still adhered to by the vast majority of Muslims spanning from the Atlantic to the Pacific.

Nasr views "Traditional Islam" as a continuous tradition that encompasses all the achievements of the 1400 years of Islamic culture. This is essential as Nasr places his interpretation of Islam in the context of current discussions over what Islam is right away. Islam, in his opinion, is not the idealized explanation favored by "fundamentalists," who demand a return to the religion's roots, the Qur'anic message in its purest form, and the Prophet's lessons, and who pursue a true form of Islam that reflects the Prophet's or the Rightfully Directed Caliphs' practice (Sunna). The statement suggests that traditional Islam is not being reinterpreted in light of modernist ideologies such as humanism, rationalism, or Marxism, which the author

[610] 'Autobiography' in: "*The Philosophy of Seyyed Hossein Nasr,*" 2001, p.85.

[611] Nasr, S.H., "Traditional Islam in the Modern World," 1985.

views as problematic. The author believes that various schools of thought, theological and juridical interpretations, sects, and orthodox and heterodox traditions, all of which make up a unified Islamic heritage, are being disregarded by both fundamentalists and modernists.

Traditional accepts in Islam:[612]

> Firstly, the Qur'an, which is believed to be the Word of God, along with its accompanying traditional commentaries known as tafsir. Secondly, the collections of Hadith from both the Sunni and Shi'i traditions, as well as traditional Hadith scholarship. Finally, Sharia refers to the body of legal rulings that have been developed and interpreted over the centuries, as well as the classical schools of law (madhab) that have emerged from them. The author acknowledges the possibility of interpreting and applying the Law to new situations through ijtihad, but only according to the classical legal principles of qiyas, ijma, and ihtihsan. The author underscores that these three sources, Qur'an, Hadith, and Sharia, are the basis for all morality in Islam, and adhere closely to the tenets of classical Islamic scholarship.[613]

According to Nasr, traditional Islam incorporates both Sunni and Shi'ite origins and interpretations as essential components. Traditional Islam is distinguished by its comprehensive scope and depth, which encompasses diverse schools of thought, theology, philosophy, art, architecture, and legal systems that constitute the entirety of Islamic intellectual heritage. Despite the multitude of manifestations, there is only one Islam, founded on the principle of unity (*tawhid*), which unites the Islamic world into a single entity. The Islamic faith is often portrayed as a distinctive tree with divine origins, with its roots firmly planted in the Qur'an and the Hadith. The trunk and branches of this tree symbolize the rich body of tradition that has emerged from those foundational roots over the past fourteen centuries, flourishing in almost every inhabited corner of the globe.[614] This Islamic world cannot be separated from Sufism. This is the deep, metaphysical, or "heart" of the Islamic revelation. Nasr suggests that the mystical element

612 Ibid., 144-15

613 Ibid., 12

614 Nasr, S.H., "Traditional Islam in the Modern World", 1985, p.11-12

of Islam is not contradictory to the external ritualistic and legal aspects of the religion, but rather serves as a complement to them. The perspective is consistent with the convictions of traditionalists and the blend of orthodoxy and mysticism championed by Al-Ghazzali, an 11th-century medieval scholar. In Nasr's perspective, the Islamic faith is inherently linked to both theory and practical orthodoxy. Consequently, a Muslim must follow ceremonial obligations, such as prayer, an annual fast, almsgiving, a trip to Mecca, food restrictions, and others, as well as restrictions regarding social, political, and economic life. Islamic art serves as a tangible representation of the spiritual essence of Islam, rendering it just as crucial to the preservation of the faith as the sharia. Along with conveying Truth, religion also embodies the concept of presence or barakah. By adhering to the principles outlined in the Qur'an, art has the potential to radiate this sense of presence.

How does traditional Islam stack up in terms of its social, political, and economic aspects? This subject is intriguing for a number of reasons, but the primary one is the political, economic, and social implications of the environmental problem, which are fundamental and unavoidable.[615] We have witnessed Nasr's promotion of the "traditional paradigm," which is designed to be a counterargument to what he sees as the defective contemporary worldview. Nasr believes that a shift in the collective paradigm would result in changes in the social, political, legal, and economic systems of countries. However, he does not provide precise descriptions of these "traditional" societal institutions in his "Perennialist" expositions. It is worth considering if Nasr provides further elaboration on the social and political consequences of traditional Islam.

In his book "Traditional Islam in the Modern World" (1985), Nasr offers specific recommendations. His proposition is that in a traditional Muslim society, sharia should be apparent in social establishments such as the family, village, and local urban neighborhoods, with the social structure being established on religious bonds. Additionally, rather than relying on impersonal and large-scale organizations, economics should be based on morality that sustains personal relationships and trust between individuals. Regarding politics, the primary goal for Muslim societies ought to be to create a more Islamic system by depending on traditional Islamic

[615] Nasr, S.H., "Reply to Lucian Stone" in: Hahn, L.E. Auxier, R.E., Stone, L.W. Jr, 'Preface' in: "The Philosophy of Seyyed Hossein Nasr," 2001, p827.

institutions. The ultimate goal is to preserve what is sacred in a perfect civilization, but there are different interpretations of this ideal.

For instance, Sunnis may support well-established establishments such as the caliphate, or in its absence, substitute political institutions such as the sultanate. These institutions have evolved over time based on the principles of *Sharia* and the needs of society. Shi'ites have access to a wide range of political institutions because they hold that the Twelfth Imam holds ultimate authority and that without him, no system of administration can be flawless. [616] Most significant is that:

> The intention is never to abolish the remains of traditional Islamic institutions that are restricted by political boundaries to create another Abu Bakr or Umar, or any kind of oppression. Additionally, such dictatorships frequently depend on external political establishments rooted in the French Revolution and other chaotic episodes in European history, despite being promoted as authentic Islamic forms of leadership.[617]

It is clear that Nasr opposes fundamentalist political ideologies as well as Islamized European revolutionary movements. Instead, he highlights the importance of the built-in limitations of traditional organizations. He believes that the political aspects of traditional Islam do not align with the intellectual and revolutionary concepts that emerged from 19th-century European philosophy. Instead, he invokes the conventional notion of socio-political revival, which is initiated by renewers (*mujaddid*), and stresses the importance of societies being revitalized from within.[618] [619]

We've noticed a growing focus on the influence and potency of a small group of spiritually awakened individuals over the broader society. These rudimentary sketches of a "traditional" Islamic socio-political philosophy are regrettably not developed by Nasr. He could be more direct in some

---

[616] Nasr, S.H., "Traditional Islam in the Modern World," 1985, p.17.

[617] Ibid., 17

[618] Nasr, S.H., "Reply to Lucian Stone" in: Hahn, L.E. Auxier, R.E., Stone, L.W. Jr, 'Preface' in: "The Philosophy of Seyyed Hossein Nasr," 2001, p.828.

[619] Nasr refers to the "great saints and sages Abd Al-Qadir Al-Jilani, Al-Ghazzali, Abu'l Hasan Al-Shadhili and Ahmad Sirhindi." In: Nasr, S.H., "Traditional Islam in the Modern World," 1985, p.17

of his previous works. This is regrettable given the profound implications of his "vision." Political instability surrounds the caliphate notion itself. A large number of modern political Islamists promote it as a societal goal.[620] In "Muslim Environmentalism," we will analyze how Fazlun Khalid, an eco-activist from the Islamic faith, merges the notion of a *caliphate* with the ecological principles of local communities. Since Khomeini created his Wiliyat-i Faqih ideology and gave the group of religious scholars political authority, Shi'ite discussions regarding the function of the state have, of course, taken on a sharper edge.[621]

Moreover, it seems that Nasr takes for granted that Islamic nations must always be established based on orthodox Islamic values. However, the question remains: is this assumption indisputable? If any nations still exist, they are few and far between in our increasingly multicultural and globalized world. Muslim civilizations have never been wholly Muslim; many of their citizens have other religious affiliations as well. What standards does he suggest be used to classify a nation as "Muslim"? Who lives in these societies who are not Muslims? What about Muslims who reside in nations that are not Muslim? What elements of the "traditional Islamic worldview" ought to serve as a model for societal transformation as a whole?

Nasr expresses concern about the fundamental paradigm or metaphysical framework that forms the basis of structures, institutions, policies, and politics. He believes that others can determine the practical implications in the domains of politics, economy, and *shari'a* law. According to Nasr, he does not possess the necessary qualifications to interpret *shari'a* doctrines since he is not a religious expert in *shari'a*. He states that his role is to uphold *shari'a* and emphasize its significance and meaning without issuing any legal decrees, as he lacks the training and authority to do so.[622]

That being said, conventional Islam is not unbiased. When criticized for being apolitical, Nasr responds: Stating that Islam is not connected to politics and is essentially apolitical would pave the way for secularism.

[620] Fuller, Graham E., "The Future of Political Islam," 2003. A categorization of various political ideologies is presented in p.47-51 en p.125-131.

[621] Nasr, S.H., "Reply to Lucian Stone" in: Hahn, L.E. Auxier, R.E., Stone, L.W. Jr, 'Preface' in: "The Philosophy of Seyyed Hossein Nasr," 2001, p829.

[622] Ibid.

Conversely, acknowledging that Islam comprises a comprehensive Sharia that governs all aspects of life, including the socio-political, personal, spiritual, and intellectual realms, without disregarding any of these dimensions, would not permit any form of secularism to seep in. Therefore, my stance aligns with the second viewpoint, not the first.[623]

Nasr, typically, abstains from engaging in politics because he is more inclined towards fundamental philosophy and natural philosophy, as opposed to philosophy regarding law, society, and politics. Additionally, he desires to remain unaffected by the turmoil of political conflict, especially in light of his personal experience during the 1979 Revolution.

However, following the events of September 11, 2001, Nasr's political involvement increased. Nasr has actively advocated for a different perception of Islam as a remedy to both modernist and fundamentalist outlooks. His objective is to demonstrate the authentic teachings of Islam in a way that Western audiences can comprehend and is in harmony with common Islamic principles. His book, The Heart of Islam: Enduring Values of Humanity (2004), serves to accomplish this goal.[624] Additionally, he actively participates in a number of interfaith and conversation initiatives, including A Common Word. We shall have to set aside this part of his life and career for the purposes of this book.

The dimension of Islam that is considered to be esoteric or inward is Sufism, which is also referred to as Islamic mysticism. According to the German scholar of Islam, Anne Marie According to Schimmel's comprehensive book "Mystical Dimensions of Islam," Sufism is the process of internalizing Islam, which involves a personal experience of encountering the central mystery of Islam, i.e., the Oneness of God (*tawhid*).

It involves a life of devoted love and the reception of immediate, unmediated knowledge from God.[625] [626] Sufism, however, is not just a spiritual path; traditional Islamic societies have greatly benefited from institutional-

---

623 Ibid.

624 Nasr, S.H. "The Heart of Islam: Enduring Values for Humanity," New York: HarperCollins, 2004. p. xiii.

625 Chittick, "The Essential Seyyed Hossein Nasr," p.74.

626 Schimmel, A.M., "Mystical Dimensions of Islam," 1975, p.3-5.

ized mysticism, such as the Sufi brotherhoods (tariqas), which are centered around a Sufi master. Despite opposition from fundamentalist organizations, Sufi-inspired movements remain significant social and political forces in Islamic cultures. It is not possible to explore this vast and fascinating universe of Sufism in detail here; therefore, we will focus on Nasr's interpretation of Sufism. As we have seen, Nasr encourages inner growth as a means of curbing consumerism (which represents the lower desires of humans) and rediscovering our sense of the sacred. What type of Sufism is Nasr advocating, given the wide range of perspectives, ideologies, and practices found in the Sufi community? What kind of religious beliefs does he follow? And more importantly, how does this impact his environmental beliefs or the other way around? With these inquiries, we are moving into a new area that is connected to traditional Islam and the Traditionalist school but simultaneously revealing a vastly different Sufi world.

Nasr's book, "The Garden of Truth:

> The Vision and Promise of Sufism, Islam's Mystical Tradition" published in 2007, is likely the most comprehensive introduction to his understanding of Sufism. Nasr's insights were developed over fifty years of scholarly study and personal engagement with Sufism. The book provides a general overview of Sufism, exploring its internal principles from the perspective of historical authorities while remaining accessible to modern Western seekers, including those who are Muslim and educated in the Western tradition.[627]

In "The Garden of Truth," the philosophical foundation of Sufism is presented, covering the essential themes of human existence, knowledge, truth (Gnosis), love, devotion, and beauty. The book also introduces spiritual ethics as a guide for human conduct and explores various spiritual practices and techniques that aid in achieving the desired state of "unity with the Divine." Sufism is recognized as the Islamic spiritual path that enables one to access the "Here and Now" – the gateway to the Divine that is always present:

> Sufism holds a distinct key that can open the gateway to our

[627] Nasr. S.H., "The Garden of Truth: The Vision and Promise of Sufism, Islam's Mystical Tradition," HarperCollins, 2007. p.xiv.

> deepest levels of being, allowing us to uncover our genuine identity, meaning, and orientation in life. Additionally, it facilitates the acquisition of profound wisdom and affection for God. Regarded as one of the world's most inclusive, well-maintained, and user-friendly spiritual pathways, Sufism offers a comprehensive system of spiritual exercises.[628]

Nasr's Sufism is undoubtedly rooted in Islamic principles, encompassing spiritual practices that are firmly grounded in the Divine Law, or Sharia, and steeped in Islamic tradition. Consequently, Nasr is uncompromising in his critique of conventional Muslim behavior, and he does not approve of pseudo-Sufi leaders and movements that have disconnected themselves from Islam, such as Gurdjieff, Inyat Khan, and Idris Shah. Sufism aims to advance towards the Truth (*haqqah*) that serves as the basis of Sharia. It commences with Sharia as the keystone of an individual's spiritual journey.[629]

Sufi metaphysics' explanation is strikingly similar to the Perennialist metaphysics of the Traditionalist school. Since each person must answer the question "Who am I?" as long as they are a human, Nasr underlines that the Sufi message is "perennial." For those whose ears are attuned to its appeal and who crave illuminating truth, Sufism's response to this age-old question still has resonance today.[630]

According to Nasr, the way of knowledge, also known as the Sufi path, involves acquiring illuminative or gnostic knowledge through spiritual practice, which he considers a vital component. He further expounds on the idea of doctrinal or theoretical Sufism (known in Arabic as *al-tasawwuf al-'ilmi*, and in Persian as *'irfn-I nazari*) or Gnosis (known in Arabic as *al-ma'rifah*, and in Persian as 'Irfn), which he perceives as the philosophical basis of Sufism. This serves as a guide to the structure of reality and the path to be taken by those seeking the Truth in order to transcend the cosmic maze. Nasr makes a distinction between practical Sufism (*al-tasawwuf al-'amal*), which involves spiritual practice and is linked to philosophia

---

[628] Ibid., 141.

[629] Ibid., 5.

[630] Nasr. S.H., "The Garden of Truth: The Vision and Promise of Sufism, Islam's Mystical Tradition," HarperCollins, 2007. p.6.

perennis, the Supreme or Sacred Science (*scientia sacra*). In contrast, he views illuminative or gnostic knowledge as part of the Sufi path, which he refers to as the way of knowledge, and as a reflection of the timeless knowledge of both the living historical tradition and Primordial Tradition.[631]

Ibn Arabi, the master from Andalusia who lived until 1224, is recognized as the initiator and innovator of the historical heritage of intellectual Sufism, also known as conceptual gnosis, during the thirteenth century. The title Shaykh al-Akbar was bestowed upon him as he was the first to develop a comprehensive overview of Sufi metaphysics or gnosis by consolidating the fundamental doctrinal components of earlier Sufism. His works had a significant impact on Sufism in the Muslim world.[632] Throughout the past eight centuries of Islamic history, there has been no scholar more significant than Ibn Arabi, with the possible exception of al-Ghazzali.[633] According to Schimmel, the importance of Ibn Arabi cannot be overstated, and Nasr agrees that his writings represent the pinnacle of mystical theories. Schimmel also categorizes his Sufism as gnostic and theosophical, drawing on the influence of Gnosticism, Hermeticism, and Neo-Platonic philosophy. Theosophical interpretations of religion often arise from the Gnostic tradition's pursuit of a deeper understanding of God and an attempt to comprehend the structure of His cosmos.[634]

Nasr followed the 'Akbarian current' and its effects on subsequent Sufism across the globe. During the fourteenth century in Persia, the Irfan tradition emerged as a blend of Ibn Arabian and Shi'ite Gnosis. [635] Kazim Assar and Allamah Tabataba'i, who were Nasr's personal tutors, follow the Irfan tradition. In the twentieth century, Shaykh al-Alawi, a Maghribi Sufi teacher, kept the oral tradition of Ibn Arabian teachings alive and introduced Schuon to the Alawiya order, with whom Nasr had a connection. The French translation of the Bezels of Wisdom by Titus Burkhardt, a Tradition-

[631] Ibid., p.32.

[632] See: Nasr, S.H. 'Theoretical Gnosis and Doctrinal Sufism and Their Significance Today' in: Transcendent Philosophy 1, 2000, p.1-36

[633] Nasr, S.H., Garden of Truth, 2007. p. 210.

[634] Schimmel, A.M, Mystical Dimensions of Islam, 1975, p.263-264. Interestingly, Schimmel qualifies Seyyed Hossein Nasr as a modern mystic.

[635] Nasr, S.H, Garden of Truth, 2007. p.185. The Persian tradition of 'Irfan is related, yet also distinct to the seventeenth century Transcendent Theosophy or Philosophy (al-hikmat al-muta'liyya) of Mulla Sadra or Sadr al-Din Shirazi (d1640).

alist, facilitated the understanding of Ibn Arabi in the West. William Chittick, a renowned specialist in Ibn Arabi's works, is Nasr's close collaborator and student. Nasr's spirituality and understanding of classical Sufism are rooted in this core, but he does not officially endorse any specific theory or practice. However, Nasr associates the Akbarian or Ibn Arabian spirit in his work with Traditionalist beliefs.

This brings up the difficult dilemma of how to understand the *theosophical* teaching of Ibn Arabi which has long been the topic of acrimonious debates in Muslim communities.[636] Schimmel asserts that there is a wide range of analyses of Ibn Arabi's work and explains this by saying that different historical influences "make Ibn Arabi's work appear quite difficult."[637] Ibn Arabi is fundamental to Nasr's thought and worldview, thus I'd want to cite Nasr extensively about what he believes to be the core ideological teachings of Ibn Arabi:

> Ibn Arabi's teaching centers around the principle of unity, which is a fundamental concept in the Qur'an. However, he expands this idea of unity beyond the oneness of God to include the notion that all of reality is essentially one. This perspective is known as the "transcendent unity or oneness of being" doctrine, which is a distinguishing feature of both Sufism in general and Ibn Arabi's school of thought. As per the transcendent unity or oneness of being doctrine, every level of existence in the universe is a reflection of a singular reality, which manifests through what Ibn Arabi and others call the "mirror of nothingness." This reflection or "theophany" (tajalli) gives rise to the whole cosmos. In line with the transcendent unity or oneness of being doctrine, Ibn Arabi delves into the nature of humankind. He elaborates that every individual, regardless of gender, possesses the capacity for all levels of existence and acts as a mirror for God to contemplate Himself. Ibn Arabi posits that the potential for the reality of the Universal or Perfect Man (al-insan as-kamil) resides within every human being, but it is only actualized in prophets and great saints of not just Islam but all genuine religions. Using these concepts as a

[636] Jong F.de, Radtke, B. Islamic Mysticism Contested: Thirteen Centuries of Controversies and Polemics, ed. Leiden, Brill, 1999.

[637] Schimmel, A.M, Mystical Dimensions of Islam, p. 265-266.

> foundation, Ibn Arabi formulates a complex system of thought that covers cosmology, sacred psychology, eschatology, epistemology, and prophetology, all unified by the wahdat al-wujud doctrine. He even explores the hidden meanings of alchemy, astrology, and other so-called occult sciences, based on the metaphysical principles he has established. Additionally, he expounds on the significance of the imaginal world and its connection to the power of creative imagination, which plays a pivotal role in spiritual life.[638]

In his discussions of the Islamic view on the organization of nature, Nasr accentuates numerous concepts, such as tawhid, *wahdat al-wujud* (Unity of Being), *walaya* (sanctity), the Perfect Man (*al-insan al-kamil*), doctrines associated with the power of language and symbolism, and the notion of the universe as a multi-layered reflection or theophany (*tajalli*). Nasr asserts that these principles serve as a foundation for authentic Islamic spirituality.

The Practice of Spirituality

The study of Sufi theory or doctrine is only one aspect of religious life, and it must be complemented by religious practice in order to achieve a complete understanding. While the study of Sufi theory may begin with the intellect, it can only become a transformative force when combined with practice, resulting in a radiant presence that transforms the entire being.[639] It is unclear what specific religious practices Nasr follows. Despite his dedication to uncovering the philosophical underpinnings of Islamic metaphysical traditions, Nasr seldom discusses his own spirituality or religious practices.[640] His memoirs reveal limited indications, such as his extensive education with Sayyid Muhammad Kazim Assar (d. 1975), Allamah Sayyid Muhammad Husayn Tabataba'i (d. 1983), and Sayyid Abu'l-Hasan Qazwini, two of whom are linked with the Persian Akbarian tradition of Irfan.

> My belief in the existence of an oral tradition of wisdom (hikmah) that could solely be attained through instruction from conven-

---

[638] Nasr, S.H., "Garden of Truth," 2007. p. 216.

[639] Ibid p.33.

[640] Nasr, S.H., "Intellectual autobiography", in: "Library of Living Philosophers," 2000.

> tional masters was influenced by various factors, such as my extensive study of Western Islamic and European medieval philosophy, direct engagements with distinguished advocates of traditional doctrines like Schuon and Burckhardt, and my childhood encounters.[641]

In the summers of 1957 and 1958, Nasr discovered his intellectual and spiritual legacy in Morocco, where he came across a strain of Sufism that was intimately connected to the spiritual heritage of the Algerian leader Shaykh Ahmad al-Alawi (1869-1934) and Shaykh Isa Nur al-Din Ahmad:

> The years spent in Morocco played a pivotal role in shaping both my intellectual and spiritual life. During this time, I found my ultimate philosophical and intellectual direction and embraced Sufism on both a theoretical and practical level. Specifically, my spiritual path became closely linked to the Maghrib and the spiritual lineage of the renowned Algerian master Shaykh Ahmad al-Alawi and Shaykh Isa Nur al-Din Ahmad.
>
> My mind and soul were deeply rooted in the world of tradition, intellectual certainty, and faith due to these experiences, which had a profound and lasting impact on me. They also guided me towards discovering inner illumination, where logic and transcendence are harmoniously united, and a blend of intellectual clarity and rigor is infused with a profound appreciation for truth and beauty.[642]

Nasr links Shaykh Ahmad al-Alawi to the Ibn Arabi tradition originating from the Maghreb, and he holds him in high esteem. In fact, he dedicated his book, The Heart of Islam, to 'The Sacred and Enduring Presence of Shaykh Ahmad ibn Mustafa al-Shadhili al-Alawi', emphasizing his respect for this master. It is important to note that Martin Lings, who is also a Traditionalist, was responsible for introducing Shaykh Ahmad al-Alawi to the Western world in 1961.[643] [644]

---

641 Ibid., p40-41.

642 Ibid., p40-41

643 See: Nasr, S.H. "Theoretical Gnosis and Doctrinal Sufism and Their Significance Today" in: "*Transcendent Philosophy*" 1, 2000, p1-36

644 Lings, M., "*A Sufi Saint of the Twentieth Century: Sheikh Ahmad al-Alawi*", Berkley and Los Angeles: University of California Press, 1961, 1974.

It's interesting to note that Nasr leaves out the fact that Frithjof Schuon's Islamic name is Shaykh Isa Nur al-Din Ahmad from his autobiography. To link the Traditionalists to this, other resources are needed.[645] This is crucial because, in 1957, Nasr was most likely inducted into the *Alawiyya* order by Schuon himself. Schuon's influence transformed this Sunni order from its basic form into a more perennial one that transcended sectarian boundaries and welcomed individuals from different backgrounds, including European converts such as Burkhardt and the Shi'ite Nasr. Eventually, the group changed its name to Maryamiyya.[646]

The idea of "Western Esotericism" The connections to Gnosticism, Hermeticism, and Neo-Platonic philosophy made by Schimmel is quite noteworthy. I would like to propose that Nasr's ideas, teachings, and practice may all be categorized as "Western Esotericism" in the way that Antoine Faivre has characterized it. In reality, Nasr himself confuses Doctrinal or Gnostic Sufism with Traditionalist teachings, which I have previously categorized as Esotericism. The hallmark of Nasr's Sufism is its commitment to the principles of correspondence, living nature, imagination, and mediation, along with the practice of personal initiation, which are the five fundamental elements of "Western Esotericism." I would want to categorize Nasr's concept of Sufism in accordance with this argument, but further evidence is required. Clearly, the traditional Islamic prophetology, which is shared by Sunnis and Shi'is, is one of the most remarkable similarities. Nasr seldom discusses it in his environmental work, although does so on occasion. It has a crucial role in Ibn Arabian philosophy. It asserts that every prophet, starting with Adam, Noah, and continuing with Moses, Jesus, and finally Muhammad, has revealed one divine Truth to humanity that is eternal and ageless. The main distinction between Islamic Perennialists and modern Western Perennialists is the latter's assertion that Islam is the most perfect expression of "Perennial wisdom" and has superseded all prior revelations. The final and final Prophet is Muhammad.[647] In the liter-

[645] Nasr, S.H., "Frithjof Schuon and the Islamic Tradition" in: Nasr, S.H., O'Brien, K, "*The Essential Sophia,* World Wisdom," 2006. Originally published in Sophia, 1999. Vol. 5. No1.

[646] Nasr, S.H., "Frithjof Schuon and the Islamic Tradition" in: Nasr, S.H., O'Brien, K, "*The Essential Sophia,* World Wisdom," 2006. Originally published in *Sophia,* 1999. Vol. 5. No1. p.260, note 2.

[647] See for example: Keller, Nuh Ha Mim, "On the validity of all religions in the

ature I've read, Nasr's viewpoint about "Islam" as the seal and abolition of all previously true faiths is not made clear. Additional study would be required. In general, his respect for other true religions does not reflect a claim that Islam is superior.

At this time, I would like to highlight a few more points. Although not widely explored, these issues are linked to several challenges that arose when engaging with Nasr's theoretical or Gnostic Sufi discourses. What is particularly intriguing about these discourses is what is left out. For instance, Nasr does not establish a connection between the Ibn Arabian legacy, Shi'ite Gnostic currents, and their integration into what he refers to as the Irfan tradition, with another significant Persian philosophical school, Shihab al-Din Suhrawardi's (d1191) School of Illumination or Ishraq tradition, which was founded by the shayk al Ishraq.[648] Nasr's earlier years saw the completion of his critical editions of Suhrawardi's complete Persian works as well as his book, The Three Muslim Sages (1963), which is the most widely translated of all his writings. This is noteworthy due to its inherent interest.[649] Ishraqi philosophy is the sole realization of the 'perennial philosophy' of humanity, as per Suhrawardi. This philosophy is thought to have originated from the divine revelations of Prophet Idris (Hermes) and transmitted through two separate channels – the Egyptian-Greek and ancient Iranian – according to Suhrawardi. It is interesting that Nasr does not examine this philosophy's connection to Persian "Akbarian" traditions given its strong resemblance to Traditionalist perennial philosophy. His motivations are unclear to me.[650]

There are a few potential factors that could be influencing Nasr's writing and the criticism he has received. Suhrawardi's historical interpretation, which has been politicized within the context of Persian "sacred nationalism," is a potential factor. This viewpoint promotes the notion of the Orient being the realm of light and the Occident being the realm of darkness. These ideas are increasingly prevalent in the Iranian Ni'matu'llahi order

thought of Ibn Al-Arabi and Emir 'Abd al_Qasir: a letter to 'Abd al-Martin", 1996.

648 Nasr, S.H, "Garden of Truth," 2007. p185; Nasr, S.H. "Theoretical Gnosis and Doctrinal Sufism and Their Significance Today" in: "Transcendent Philosophy" 1, 2000, p.1-36.

649 Izutzu, Toshihiko, "Ishraqiyya" in: "Encyclopedia of Religion", second edition, p.4552-4556.

650 See "Ishraiyya" in: "Encyclopaedia of Religion."

under the leadership of Javad Nurbaksh.[651] It's possible that Nasr wants to distance himself from this nationalist discussion, as he hasn't explicitly referenced Persian nationalism in his writing despite his deep involvement with Ni'matu'llahi and his lifelong focus on Persian mysticism.[652] However, Nasr has faced criticism for his supposed "confessional esotericism" and "Islamic apologetics." There are accusations that he is using these ideas to advance his own personal, ethnic, or religious chauvinistic agenda, and some say that he has created a partial historical account or even a "Nasr School" of Persian philosophy. In response to these criticisms, Nasr has delved into the connection between Western scholarship and its historical analysis and conventional Islamic scholarship.[653] [654] Regardless of whether or not this critique is valid, it's worth considering the potential influences on Nasr's work and the criticism he has faced.

I would also want to draw attention to the remarkable similarities between Nasr's views of Islam and mysticism and Ismaili beliefs. Nasr's alleged Ismaili heritage has given rise to considerable controversy in Turkey, according to some.[655] Nasr's expositions on this topic make it very evident that he sympathizes with Ismaili's teachings. In An Anthology of Philosophy in Persia (2005), Nasr expounds on the Hermetico-Pythagorean and Ismaili philosophies, which he describes as embodying the conventional and time-honored sense of philosophy and being closely linked to the esoteric (*batini*) principles of religion. Nasr considers these philosophies to be genuine examples of timeless perennial wisdom. Of particular interest to Nasr are the Rasa'il Ikhwan as-Safa (Treatises of the Brethren of Purity), which he finds particularly intriguing. Nasr emphasizes this text in a number of his books since it is one of the most important to Ismailis.[656]

---

[651] See the discussion about the "Perzianization of Sufism" by Leonard Lewisohn in his article 'Persian Sufism in the contemporary west: reflections on the Ni'imatu'llahi diaspora' in: Malik, J, Hinnells, J, "Sufism in the West", Londin, New York: Routledge, 2006.

[652] L. (ed.), "The Heritage of Sufism," Oxford: Oneworld, 1999.

[653] John Walbridge, The Wisdom of the Mystic East: Suhrawardi and Platonic Orientalism, 2001.

[654] Nasr, S.H. 'Introduction' in: Encyclopaedia of Islamic Spirituality, volume One Foundations, 2000.

[655] Edis, T., An Illusion of Harmony, Science and Religion in Islam, New York: Prometheus Books, 2007. p.34.

[656] .H. Nasr., An Introduction to Islamic Cosmological Doctrines, Albany: State Uni-

> The Treatises of the Brethren of Purity stand out for their focus on not just highlighting the concealed nature of true philosophy, the different stages of initiation and levels of knowledge, and the merging of philosophy with spiritual attainment and moral behavior that are characteristic of Ismaili philosophy, but also for their clear elucidation of Islamic Pythagoreanism and Hermeticism.[657]

Moreover, Nasr emphasizes the significance of a specific Ikhwan as-Safa treatise, namely, "The Debate between Man and the Animals," in relation to the environmental problem. Muslim environmentalists have also referenced this text as evidence of Islamic ecological philosophy.

> During a time when humans are infringing upon the rights of other creatures and devastating the environment with the belief in their complete dominance over creation, Ikhwan's philosophical reasoning regarding animal rights is highly pertinent. Their ideas showcase an "ecological philosophy" that is of great importance in the development of Islamic environmental philosophy and in addressing urgent ecological concerns.[658]

Traditionalism

Nasr frequently describes himself as a "traditionalist" (occasionally written "Traditionalist"), and anybody who is familiar with his writings will quickly find him to be categorized in this way. To avoid misinterpretation and misunderstanding, however, the term "traditionalist" has to be defined carefully and mindfully. For instance, a "traditionalist" might signify many things depending on the setting of Islamic studies. It may be useful to identify a group of individuals who placed a high value on the Traditions, in this case, the Hadith, which are compilations of sayings and acts ascribed to the Prophet, in historical Islam study. It could also refer to organizations that follow regional Islamic "traditions" or cultures.

There are some who argue that the emphasis on traditionalism, which

versity of New York Press, revised edition 1993. First edition 1964.

657 Nasr, S.H., Amanirazavi, M. (eds) An Anthology of Philosophy of Persia, 2005. p.11

658 Ibid.

prioritizes the reliance on tradition as the primary source of authority, is a notable characteristic of Islam as a complete way of life. Nasr is undoubtedly a traditionalist in this sense. His understanding of "tradition" differs from others, who in the realm of Islam are more commonly referred to as "traditionalists."

If one does not recognize the connection between Nasr's traditionalism and the school of thought known as Traditionalism, one will not be able to fully appreciate Nasr's work. Traditionalism, which is also referred to as perennial philosophy, is a Western philosophy that originated from the works of Frithjof Schuon (1907-98), Ananda Coomaraswamy (1877-1947), Réné Guénon (1886-1951), Titus Burckhardt, Martin Lings, and Marco Pallis. According to a biography of Nasr, his intellectual and spiritual journey was greatly influenced by the discovery of traditional metaphysics, philosophia perennis, and the writings of Frithjof Schuon. These writings emphasized the need for spiritual discipline in addition to theoretical knowledge.[659]

The work of Coomeraswamy, Guénon, and Schuon influenced a number of academics, including Nasr, who began to routinely publish in a variety of European journals in the 1950s and 1960s. The scholars quickly became known as the "Traditionalist school," known for their critical examination of modernity and profound admiration for classical metaphysics, cosmology, and philosophy from a range of religious traditions. Traditionalism, as described in the book "Traditionalism: Religion in the Light of the Perennial Philosophy" (2000), aims to reveal the shared metaphysical foundation of all religions, known as the philosophia perennis.[660]

> By definition, the Traditionalists are devoted to elucidating the philosophia perennis, which forms the essence of various religions and exists beyond the diverse manifestations of the world's distinct traditions. Moreover, they are committed to safeguarding and illuminating the conventional structures that lend each religious legacy its purpose and secure its formal consistency, thereby ensuring its spiritual potency.[661]

---

[659] "Biography of the website of the Nasr foundation". www.nasrfoundation.org

[660] Oldmeadow, K., *"Traditionalism: Religion in the Light of the Perennial Philosophy,"* Shrilanka Institute of Traditional Studies, 2000.

[661] Ibid p.viii.

Transcendent Unity of Religions

At the core of Traditionalist philosophy is the concept of the "transcendent unity of religions." This idea maintains that all major traditional religions fundamentally reflect a single ultimate Reality. Despite their differences and diversity, there is an eternal and unchanging truth, a transcendent unity that exists at the heart of all religions. According to Traditionalists, the major world religions are expressions of this ultimate oneness, with timeless and enduring shared foundations. They believe that authentic or integral religions are grounded in a specific "revelation" intended for a particular human society at a particular time. For instance, Christianity's foundation is rooted in the revelation of Christ, while Islam's is based on the Quran. The term "tradition" encompasses the original revelation as well as the evolving and developing nature of that revelation across different cultures and time periods. This encompasses fields such as metaphysics, cosmology, philosophy, religion, social structures, arts, and architecture. Traditionalists view all of these diverse traditions as expressions or echoes of a single Primordial Tradition, representing humanity's spiritual and intellectual heritage. This legacy was acquired through direct revelation when heaven and earth were still connected, and it serves as the original and archetypal foundation of human knowledge.[662]

Traditionalists believe that since all traditions originate from the same divine source, they share similar perspectives on the significance of human life and the fundamental aspects of human cognition, despite existing in vastly different cultures, such as those of the Eskimos and the Australian Aborigines, as well as the Taoists and Muslims. As a result, it is conceivable, according to Traditionalists, to develop a comparative theology or metaphysics of religions. The intellect, also known as the eye of the heart, is a uniquely human faculty and a constituent of the soul that can perceive the Truth of eternal knowledge. This differs from discursive thinking, which is commonly linked with the intellect and requires a more profound level of comprehension or reflection.

---

[662] Nasr, S.H., "The Traditionalist Approach to Religion", in: Chittick (eds), "*The Essential Seyyed Hossein Nasr*", p.21.

Esoteric, Exoteric and Orthodoxy in Practice

The philosophy of Traditionalism is characterized by religious orthodoxy, both in doctrines and rituals. Although there could be multiple routes to attain the absolute truth, Traditionalists stress that this does not imply that we can select and adopt various religions merely for our convenience. Instead, humans must choose one of the pathways that God has deemed fit for us, as outlined by integral and authentic religions. Without choosing a specific path, there is no way to access the spirit, which requires adherence to the beliefs and practices of a particular religion's theological orthodoxy. This orthodoxy mainly focuses on the exoteric aspects of religion, including specific religious teachings and practices.[663] The esoteric aspects, in contrast, allude to their internal purpose and significance. According to the Traditionalists, esotericism cannot exist without conventional exotericism; otherwise, esotericism will turn into occultism. The majority of Traditionalists have harsh opinions of contemporary eclectic and 'occultist' spiritual practices. We already know how harsh Nasr is about New Age movements and their pseudoscience.

Traditionalist ideology is fundamentally based on this gap between the esoteric and exoteric. It could make it easier for us to comprehend why religions aren't all same while having similar underlying beliefs, as several critics have suggested. Religions vary in form, organization, theory, and practice as they are manifested in the realities of daily life. High-abstract terminology prevails in perennial literature. It frequently lacks a clear explanation and is challenging for readers who are not familiar with it to access (like myself). In an effort to clarify the Traditionalist approach to religion, Nasr himself offers the following example:

> The philosophy of perennialism posits that reality is not limited to the psychophysical world that humans typically operate within and that consciousness goes beyond the level of awareness commonly experienced by present-day humanity. The ultimate reality is unbounded and absolute, generating goodness akin to how the sun's rays emanate from it. The Pure Object principle can also be viewed as the Pure Subject or the Supreme 'I', wherein regular

[663] Nasr, S.H., "Religion and the Environmental Crisis" in: Chittick, *"The Essential Seyyed Hossein Nasr"*, p. 29

> consciousness is considered an external covering of the Supreme Self, rather than an expression of the Supreme Reality within the lower realm of the universal hierarchy.
>
> Religion is not merely the convictions and customs of a specific human society that has received a particular religious communication. It is rather a heavenly truth whose prototype stems from the Divine Intellect and has levels of significance and existence akin to the universe itself. In the Platonic understanding, religion is an Idea that endures in the Divine Intellect, having a reality beyond time.[664]

In order to fully understand the Traditionalist ideas and teachings, we would need to engage in a specialist theological conversation. However, when reviewing Nasr's environmental philosophy in the prior Chapter, we already saw a rough summary of some of the key ideas and principles. It includes the conviction that God is at the center of reality, that existence is one, that the cosmos is ordered in a hierarchy, that man occupies the center of reality, that nature and man are intimately connected on all planes of reality, that there is a sacred science, and that primordial Truth exists. This concept, when applied to the environmental catastrophe, might be categorized as a particularly "Traditionalist" or "Perennialist" approach. We'll see whether other Traditionalists adopt this eco-traditionalist perspective.

It is not possible to fully explain Traditionalist metaphysics here. It is advised for people with an interest in the topic to read *Traditionalism: Religion in the Light of the Perennial Philosophy* (2000). Understanding the idea of the transcendent unity of all religions is essential to grasp Nasr's belief that studying multiple traditions can help us uncover a sacred science governing the natural world. The strong commitment to orthodoxy, which refers to conformity to religious ideas, practice, and legislation, is also relevant. Since no two religions are the same, it is imperative to follow one of the recognized orthodox religions.

---

664 Nasr, S.H., "The Traditionalist Approach to Religion", in: Chittick (eds), "*The Essential Seyyed Hossein Nasr*", p.21.

Perennial Philosophy

The concept that all faiths share a similar perennial philosophy—a timeless truth—is not exclusive to Traditionalists. Agostino Steucho, a Vatican librarian in the sixteenth century, is really attributed with coining the phrase "perennial philosophy" when he wrote *De Perenni Philosophia* in 1540. However, the idea dates back much further, to the Latin and Greek Church Fathers as well as Philo of Alexandria's philosophic theology. The idea of a single ancient philosophy and theology dating back to paradise, as put forth by subject expert Wilhelm Smidt-Biggemann, "has its roots in late antiquity, especially with the Christian church fathers, and was newly adopted by the Florentine philosopher Giovanni Picodela Mirandola who called it *philosophia prisca*."[665] During the Renaissance, Nicholas of Cusa, Marsilio Ficino, Giovanni Pico della Mirandola, and Agostino Steuco all embraced the concept of the perennial philosophy, and subsequent thinkers followed in their footsteps. Rediscovering Hermeticism and Neo-Platonism was necessary for this. In the seventeenth century, Enlightenment figures like Leibniz, Lessing, Voltaire, and others also adopted the idea of perennial philosophy. Often, this meant recognizing that perennial knowledge had been passed down through a line of great teachers, beginning with figures such as Zoroaster, Hermes, Orpheus, Plato, Jesus, and others.

Nasr holds a skeptical standpoint regarding the humanist interpretation of Renaissance thinkers like Ficino and Pico. He contends that they played a significant role in formulating a distinct notion of freedom, namely, the inherent liberty of humanity from all limitations. Nasr posits that this concept is the fundamental cause of the present dilemma. Conversely, Nasr values the perspectives of Neo-Platonism and Western Hermeticism, which are genuine Western legacies.[666] Nonetheless, Nasr staunchly objects to the occult movements that materialized during the 19th and 20th centuries, including Gurdjieff, the Order of the Golden Dawn, and Blavatsky's Theosophy. Moreover, he scrutinizes contemporary spiritual or esoteric groups that espouse the concept of universal harmony among all religions. The well-known book, Perennial Philosophy by Aldous Huxley, which is

---

665 Schmidt-Biggeman, W., "*Philosophia Perennis: Historical Outlines of Western Spirituality in Ancient, Medieval and Early Modern Thought,*" International Archives of the History of Ideas, Dordrecht: Springer, 2004.

666 Nasr, S.H., "*Religion and the Order of Nature,*" 1996, p131, p173

frequently censured by Traditionalists, is where most individuals nowadays are first exposed to the term "Perennial Philosophy."[667]

A lot of misinterpretation has been caused by the idea that all philosophers and movements that support a "perennial philosophy" can be grouped together under this flag: traditionalists like Coomeraswany and Schuon are herded together with syncretists, pseudo-mystical romantics, theosophists, anthroposophists, Gurdjieffians, neo-Deists, neo-Hindu universalists, and the occult.[668] These movements and ideas are referred to as "pseudo- or counter-traditions." The significance of religion's role and purpose distinguishes traditionalists from progressives. Religions that are "genuine" in the eyes of Traditionalists are those that are founded on revelation. These are "self-sufficient and completely adequate in giving all aspects necessary to the civilizations in which they have emerged."[669] More crucially, there might be no compromising when it comes to the exterior, or "exoteric," manifestations of revealed religious practice. Orthodoxy, or following the rules of the faith, is a must. A Christian traditionalist must be orthodox, while a Muslim traditionalist must follow the *sharia'ite* laws, which include the five everyday prayers, the Ramadan fast, the Mecca pilgrimage, giving alms (zakat), and other social laws.

The relationship between a person and God is at the foundation of all religions, according to Nasr. Only one part of this relationship is highlighted by each religion, yet regardless of the restrictions imposed by its many forms, each one's principles ultimately include the truth itself. Because of this, fully embracing one religion entails fully embracing all religions. Trying to create a universal blend of different religions while actually dismantling the established forms that connect people to God is a meaningless and harmful action. This act undermines the only means by which individuals can develop a genuine attachment to God.[670]

Although Traditionalists are often regarded as advocates of a timeless

---

[667] Huxley, A., "*The Perennial Philosophy,* Harper &Row", New York: 1970.

[668] Oldmeadow, K., "*Traditionalism: Religion in the Light of the Perennial Philosophy*", Shrilanka Institute of Traditional Studies, 2000. p158-159

[669] Oldmeadow, K., "*Traditionalism: Religion in the Light of the Perennial Philosophy,*" 2000. p.144.

[670] Nasr, S.H., "*Islam and the Realities of Islam,*" Allan & Unwin, 1975, p16.

philosophy, they differ from other schools of thought that employ the same concept. Therefore, while the Traditionalist viewpoint may be centered on perennial philosophy, it is important to note that their beliefs are unique. According to Oldmeadow's concise assertion, the matter of religious institutions acts as a yardstick to differentiate Traditionalists from other alleged perennial philosophers who tend to react defensively upon hearing words like dogmas or orthodoxy.[671]

For a thorough comprehension of Traditionalist and Perennialist philosophies, it is necessary to embark on a journey into the realm of "Western esotericism," a term coined by Antoine Faivre, a prominent scholar of esotericism. Faivre's elucidation of Western esotericism will aid us not only in classifying Nasr's viewpoint but also in connecting his ideology with the contemporary religious and spiritual movements of the New Age and ecological movements. Therefore, it is imperative to include Faivre's explanation in our analysis:[672]

Faivre formulated a systematic definition of esotericism as a particular worldview or method of comprehending the universe, nature, and humanity. He identified six components that characterize this unique form of thought, four of which are fundamental to esotericism while the remaining two are context-dependent. In view of its importance to the inquiry, I will cite Kocku von Stuckrad's words from the Encyclopedia of Religion and Nature:[673]

1. *The tenet of correspondences postulates that every element of the perceptible and imperceptible cosmos is linked and symbolically reflects one another. This principle is observable in the notion of microcosm and macrocosm, where there exist correlations among various strata of the physical and metaphysical realm, as well as among nature, history, and revealed scriptures.*

2. *The concept of a living nature doctrine sees the universe as a dynamic system where all components are interconnected and animated. Its*

[671] Oldmeadow, K., *"Traditionalism: Religion in the Light of the Perennial Philosophy"*, 2000. p.145

[672] Antoine Faivre held the first chair for "History of esotericism and Mystical Current in Modern and Contemporary Europe" (Sorbonne, Paris).

[673] Stuckrad, K. "Western esotericism" in: Taylor, B. *"Encyclopedia of Religion and Nature"*, 2 volumes, New York: Continuum, 2005, 2008. p.1728-1729.

origins can be traced back to both Western philosophy and ancient and Renaissance magic.

3. *The idea of imagination and mediation proposes that there is a potential for connection between the higher and lower worlds through ritual, symbolic acts, or intermediate spirits such as angels. Imagination is regarded not just as a tool for focusing during magical rituals, but also as an "organ of the soul" that creates a cognitive and visionary bond with the intermediary realm.*

4. *The notion of transmutation experience relates to an individual's personal initiation on a spiritual journey through either a private change of status or a social event.*

In addition to these four fundamental components, two more features are relevant:

5. *The praxis of concordance, which aims to demonstrate commonalities between different traditions, often through the belief in a prisca theologia or philosophia perennis that captivated medieval and Renaissance scholars.*

6. *The transfer of esoteric teachings from a master to a disciple highlights the significance of being introduced to a spiritual tradition and the validation of authority or "authenticity" through an unbroken lineage of ancient masters.*

According to historical evidence, the revival of Neo-Platonism and Hermeticism in fifteenth-century Europe marked the beginning of Western esotericism. This resurgence, as noted by Faivre, involved the restoration of a traditional cosmology, which served as a counterbalance to the increasing secularization and mechanization of the world.[674]

> Two consequences arose from the realization that nature is a cohesive and regulated domain: the secularization of the universe, which decreased the respect for the divine, and a resurgence of magia, characterized as a collaborative philosophy of nature.

---

[674] References to: Faivre, A. "Ancient and medieval Sources , Esotericism' and 'Renaissance Hermeticism"

Wouter Hanegraaff, a Dutch academic, explains that a trend in the eighteenth and nineteenth centuries gave rise to two separate streams: occultism (e.g., Blavatsky's theosophy) and Romanticism. These two streams can be seen as responses to a clash of worldviews and efforts to modernize traditional ideas. According to Hanegraaff's book, "New Age Religions and Western Culture – Esotericism in the Mirror of Secular Thought" (1996), occultism, which has a stronger link with modern science than Romanticism, re-emerged as a New Age religion during the 1980s and 1990s. Hanegraaff examines and documents this development of New Age religions, stating that the New Age can be viewed as a "secularized" form of esotericism. Hanegraaff notes that both esotericism and Christian theology have attempted to modernize traditional beliefs to adapt to a secular world created by secularization.

> Christianity and esotericism have both created religious theories, ideas, and practices that exist somewhere between completely secular mindsets and defenders of traditional beliefs against modern influences. This type of religion is seen as too religious for those with secular beliefs and too secular for those with traditional beliefs. In both Christian theology and esotericism, there are traditionalist conservatives and modernist liberals. In the case of esotericism, for instance, the conservative group, exemplified by Guenon, condemns occultism as a distortion.[675]

I suggest that Nasr's Traditionalist thought could be categorized as an expression of non-secular Western Esotericism, similar to Traditionalist teachings. This aligns with Faivre's description of Esotericism. Nasr's opposition to secular modernity and his emphasis on traditional models make Hanegraaff's book title, "Esotericism in the Mirror of Secular Thought," relevant to his work. In my opinion, to fully understand Traditionalism and its distinct understanding of perennial philosophy, we must view it within the broader context of movements and philosophies emerging in the Western world, falling under Faivre's concept of Western Esotericism. Given Nasr's deep commitment to his Islamic heritage, it is essential to analyze his views within an Islamic context. There is a debate about whether Nasr's Traditionalist perspective of Islam can be considered a form of Western

[675] Hanegraaff, W. *"New Age Religions and Western Culture – Esotericism in the Mirror of Secular Thought,"* Leiden: Brill, 1996. p.408

Esotericism, which will be explored shortly. If it is the case, why use a term such as 'Western Esotericism,' which is specific to the West? Wouldn't it be more appropriate to use a broader category of Esotericism? Proponents of perennial philosophy argue that esotericism constitutes the innermost essence of all valid religions and thus has universal significance. Various scholars have observed the apparent "universality" of spiritual, esoteric, or "gnostic" movements, and have offered different explanations for this occurrence.[676]

Perennialists often encounter criticism regarding the notion of universality and the existence of timeless knowledge. Some argue that the idea of universal wisdom is a product of Western thinking, which imposes Western concepts onto Eastern traditions. Hanegraaff highlights that in the 19th century, there was a tendency to romanticize and adapt Oriental religions to fit within Western contexts.

> Despite the widespread fascination with the East, Westerners living in the late 19th century struggled to gain a comprehensive and impartial understanding of how Hinduism or Buddhism operated within their own cultural milieu.[677]

Can the identical be stated of the early Traditionalists of the 20th century, like Schuon, Guénon, and Burckhardt, some of whom played a key role in introducing Islamic 'esotericism' to the West and had inspired Nasr? They were all Europeans, and their efforts to "create" a Westernized understanding of Islam have drawn criticism. Nasr answers these accusations and is aware of these disputes. He firmly believes that Frithjof Schuon has presented the most intrinsic element of the Islamic message to Western readers without tailoring it to suit European perspectives. Despite receiving his education in the West, Nasr also received instruction from some of the most eminent personalities in Persian spiritual and philosophical traditions. Therefore, it cannot be argued that he is creating a Westernized version of Islam.[678] As

---

676 In: "The Science of religion in Israel" in: G.A. Wiegers, J.G Platvoet (eds.), "Modern Societies & the Science of Religions" (NUMEN Book series 95), Leiden 2002, p.258-271, p.268-270.

677 Hanegraaff, W. "New Age Religions and Western Culture – Esotericism in the Mirror of secular Thought", Leiden: Brill, 1996. p.462

678 Nasr, S.H., *"Knowledge and the Sacred"*, Albany: State University of New York Press, 1989. p.67

we will see, several critics claim that he does. We must first talk about the brand Islam Nasr envisions. We'll find that his Islamic esoteric ideas have striking similarities to Faivre's definition of esotericism.

It is interesting to note that the majority of the founding figures of traditionalism, including Réné Guénon, Titus Burkhardt, Frithjof Schuon, and Martin Lings, were scholars of Islam who often converted to Islam. It is not usually acknowledged that traditionalism has a uniquely Islamic component. From its root, several branches emerged. We shall examine this later when analyzing modern Traditionalist networks. Some were to highlight Islam more, others less. For instance, Schuon for the first time formed the Western branch of the Sufi order known as the Alawiyya, which eventually changed its name to the Maryamiyya. Guenon later rose to prominence as a renowned sheikh in Egypt. Born a Muslim, Nasr was particularly influenced by Frithjof Schuon's 'Perennialist' interpretation of Islam. Many additional aspects of Nasr and Schuon's relationship will be explained. Nasr describes the importance of these folks to his intellectual development in his autobiography:

> My intellectual growth was significantly influenced by my encounters with distinguished representatives of the perennial philosophy and tradition in Europe, starting in 1957. Among these figures, Frithjof Schuon was particularly noteworthy and I maintained a close association with him until his passing in 1998. Additionally, Titus Burckhardt had a significant impact on my thinking, particularly in the areas of traditional cosmology and the philosophy of art. Marco Pallis introduced me to the metaphysics of Tibetan Buddhism, while Martin Lings, a key exponent of traditional Islam and Sufism in the Western world, has been a close associate of mine for the past four decades. The impact of these individuals, including Rene Guenon and A.K. Coomaraswamy, on my intellectual development cannot be emphasized enough.[679]

[679] "An Intellectual Autobiography" in: Hahn, L.E. (eds), "*The Philosophy of Seyyed Hossein Nasr*," 2001, p.40-41.

Chapter Summary

Nasr's ecological ethics is deeply rooted in traditional Islam and draws on the Qur'an, Hadith, and Sharia for its ethical principles. He argues that Islamic teachings have a rich tradition of ecological ethics that can inform contemporary environmental thinking. Nasr highlights the importance of the concept of *haqq*, or the right of every living being in accordance with its nature. He also draws attention to the Prophet's teachings and acts, which encourage the protection of nature and animals, and the communal management of natural resources. The Sharia provides both specific regulations and overarching principles for addressing ecological crises, including restrictions on usury, extravagant consumerism, and excessive accumulation of wealth. Nasr's Islamic environmental teachings have the potential to create significant structural and systemic changes in the current global economic and financial institutions that prioritize wealth growth. Ultimately, Nasr's perspective emphasizes the importance of a deep understanding of nature for creating effective environmental laws, and how a profound reverence for nature is essential in environmental thinking.

The final chapter summarizes Nasr's key ideas and their implications for the environmental movement, including how his emphasis on the sacred dimension of nature can inspire action to protect the earth. The chapter also discusses potential challenges and benefits in implementing Nasr's suggestions and the need to restore the sacred dimension of nature for a sustainable relationship with the natural world.

Chapter 5

The Meanings and Influences of Nasr's Ecological Ethics

As we come to the end of this book, we are left with a deeper understanding of Nasr's suggestions regarding the contemporary environmental crisis. Nasr's work has highlighted the importance of traditional metaphysical systems in addressing this crisis and has emphasized the need to recognize the sacred nature of the natural world. Through his writing, Nasr has challenged the reductionist and exploitative perspective of modern society and offered an alternative approach to environmental issues.

In this final chapter, we will summarize Nasr's key ideas and their implications for the environmental movement. We will explore how his emphasis on the sacred dimension of nature can inspire individuals and communities to take action to protect the earth. We will also discuss the challenges that may arise in implementing Nasr's suggestions and the potential benefits that could be gained from embracing them.

As we look to the future, the environmental crisis continues to pose a significant challenge to humanity. However, through the insights offered by Nasr's work, we can gain a greater appreciation for the intrinsic value of nature and the need to recognize its sacred nature. By embracing traditional metaphysical systems and restoring the sacred dimension of nature, we can take steps towards a more sustainable and harmonious relationship with the natural world.

Traditional Metaphysics

It is reasonable to advocate for a significant change to the common modern perspective due to its inherent shortcomings. The current modernist viewpoint has resulted in limited understanding and a narrow, incomplete perception of reality. A new comprehensive perspective is necessary, including new beliefs and theories about the nature of existence. Nasr ardently calls for a critical reevaluation of our shared worldview, including our conceptions of what it is to be human, what the environment is, and the place of mankind in the "greater scheme of things." According to Nasr, "this radical transformation paradoxically requires not a new discovery, but the rediscovery of our relation to nature, which has always been evident to a large extent in the life of traditional societies over the ages."[680] The moment has arrived to recover this "traditional wisdom," which once represented fundamental truths about the nature of existence.

> A comprehensive approach to our understanding of the world requires a combination of various disciplines such as religion, art, philosophy, mysticism, and sacred science. Along with these traditional forms, a new philosophy of nature is needed which is an updated version of the perennial philosophy of nature.
>
> To achieve this goal, non-Western religions, including primal ones and Islam, must be recognized and given a prominent role. Additionally, it is crucial to reviving the Western spiritual tradition regarding nature, while reformulating the traditions of other cultures before they too fall into oblivion as the Western tradition did.[681]

Nasr contends that religion is fundamental to tackling the environmental crisis since traditional paradigms have their roots in religion. He maintains that major world religions have always focused on comprehending the truth of reality and nature, but modern society has disregarded this aspect. According to Nasr, the tragedy lies in the fact that modern society has excluded religion and its associated knowledge of the truth, leading to a disconnection from the ancient wisdom that is crucial to understanding the environment. To address the environmental crisis, humanity must rediscover the timeless knowledge about nature and the Earth embedded within our cultural traditions. It is imperative to rekindle and comprehend the sacred science of the order of nature to pave the way for a sustainable future.

Nasr's philosophy centers around the idea of "sacred science," which refers to knowledge that is eternal, universal, and rooted in the divine order of reality. Later chapters will delve into Nasr's ideas on Perennial philosophy and Traditionalism, exploring the concept, also known as perennial

---

680 *Sophia.*, 2004, p..6.

681 *Sophia*, 2004, p.12.

wisdom. Nasr's book Religion and the Order of Nature aims to reveal the sacred understanding of nature that forms the core of all authentic religions, despite their diverse forms. It highlights the common foundation and key principles shared by all religions.

Traditional knowledge framework is defined by five basic principles:

> (1) Belief in the existence of a single divine absolute truth suggests a monotheistic belief in a singular ultimate reality. This belief is common in many religious traditions. (2) Recognition of the hierarchical order of the universe refers to the understanding that there are different levels of complexity and organization in the cosmos. (3) Acceptance of the existence of cosmic laws that govern the universe acknowledges that there are fundamental principles and rules that underlie the workings of the universe. (4) Acknowledgement of the inherent unity of existence recognizes that everything in the universe is interconnected and interdependent. (5) Recognition of the central position of humanity suggests that humans have a unique and significant role to play in the universe.

Nasr posits that the primary principle of reality is the presence of a single, objective, and absolute truth, which is heavenly in nature. However, this truth is not acknowledged by post-modernists who are generally skeptical of absolute truths.[682] Despite this, Nasr contends that the post-modern paradigm, while rejecting absolutes, still places humans at the center of reality without question. In contrast, Nasr's worldview centers on God and regards all other characteristics of reality as mere manifestations of the heavenly order. Nasr further asserts that authentic religions reflect this one reality and challenge the anthropocentric vision of the modern prototype.

The second principle concerns the hierarchical order of reality, with levels of existence arranged in a hierarchy that spans from the material realm at the bottom to the divine realm at the top, known as the "great chain of being." [683] Each level reflects the perfection of the divine to a greater degree. As

---

682 *Sophia,* 2004, p.11.

683 The term "Great Chain of Being" is occasionally used by Nasr. It is used in a work by A. O. Lovejoy, *"The Great Chain of Being,"* Cambridge: Harvard University Press, 1913. Lovejoy traced the concept of this type of idea – of a hierarchical interconnected worldview - through western history.

humans, our perception and understanding of the physical world are limited to our senses and observations, but there are many other levels of reality beyond this. The physical world is linked to God through these higher levels of reality that make up the various stages of the cosmic hierarchy.[684]

The third principle, known as 'cosmic law', asserts that the reality we inhabit is structured and regulated by laws and principles at every level, encompassing the physical, psychological, and spiritual realms. These laws are observable in the material world, such as the law of gravity governing the descent of an apple, and form the basis of physics. While less apparent in other domains, psychological and spiritual realities are similarly subject to universal laws, such as the principle of cause and effect in karma.

The fourth principle highlights the interconnectedness and interrelatedness of all levels of reality, as they reflect one divine reality. This concept called the inherent unity of existence, proposes that humanity and nature are intricately linked in ways that we have yet to fully comprehend:

> We are interconnected within the complex fabric of life in a way that cannot be simplified to mere quantitative or material relationships. It is crucial that we recognize and reembrace the perspective of nature as a magical and interconnected reality, to which we have a connection that extends beyond the physical, chemical, and biological levels, and includes the psychological, intellectual, and spiritual aspects.[685]
>
> The branches of nature are our branches, her existence is our existence, and her demise is our demise. Despite formal differences, all religions share the belief that the natural order and the human order are connected and are both marked by the Divine reality, which is the source of both humanity and nature.[686]

Man is seen as occupying a central position in the interconnected, enchanted universe described as a hierarchical "vertically ordered" structure.[687] In

---

684 *Sophia*, 2004, p.12.

685 *Sophia*, 2004, p.12.

686 Nasr, S.H., *Religion and the Order of Nature*, 1996, p.24-25.

687 *Sophia*, 2004, p.11.

the realm of traditional metaphysics, it is believed that humans serve as a bridge connecting the divine realms above and the terrestrial realms below. They are viewed as the intermediary between Heaven and Earth. The constitution of humans, comprising their body, psyche/soul, and spirit, mirrors the various levels of the cosmic hierarchy. Their physical body is linked to the material realm, their psyche/soul to the realms of imagination, and their spirit to the higher spiritual planes, ultimately leading to the divine. Humans are believed to be designed to progress from lower to higher levels, with the microcosm of man being related to the macrocosm of the universe. Religious beliefs suggest that man is created in God's image, representing Divine qualities. Despite his central position, man's power is restricted by the laws that govern the universe. He can dominate and harm the natural world but is subject to physical, psychological, and spiritual laws. In the end, humanity possesses restricted rights and obligations in regard to the act of creation.[688]

> Upon rekindling our understanding of humanity's authentic essence, we shall acknowledge that solely the Ultimate or Heaven holds absolute status. Despite being a reflection of the Absolute and designed to pursue it, humans reside in this world as creatures with relative rights that come with obligations. Consequently, humans cannot exercise dominance over and destroy their surroundings in the name of the absolute nature of humanity and its alleged "requirements."[689]

> Our rights are restricted by acknowledging the rights of other beings, and our responsibilities expand beyond our family, tribe, nation, or religion to include all of humanity and nature.[690]

Humans possess the ability to make choices freely, however, this liberty is constrained by the laws that govern the cosmos. Violating these regulations leads to imbalance and disrupts cosmic harmony. While individuals may choose to pollute oceans, they must also accept the consequences, such as poisoned food and potential harm to their health. These examples demonstrate the law of cause and effect in the physical realm. However,

---

688 *Sophia*, 2004, p.11.

689 *Sophia* 2004, p.11.

690 *Sophia* 2004, p.10.

this concept often referred to as "karmic law," extends to other areas as well. It is crucial for humans to respect these laws to maintain balance and equilibrium. The current environmental crisis is a result of disregarding or disregarding these principles, and the natural order of things.

This traditional viewpoint as described here has some significant implications that also have an impact on how we relate to nature.

First, In traditional worldviews, morality, ethics, and the laws of the universe are interconnected. This can be difficult for modern minds to understand, as they are used to separating morality from reality. However, this interconnectedness is crucial to Nasr's traditionalist ideas. According to Nasr, human liberty is restricted by universal laws, and this liberty must be balanced with restrictions and responsibility. Without this balance, chaos and imbalance result. Nasr believes that society currently lacks this sense of moral restraint and that it must be rediscovered and reinforced through the study of traditional knowledge and spiritual training. Without this metaphysical moral order, effective environmental ethics is not possible.

> It is a tragic reality that the prevailing way of thinking has separated ethics from metaphysics and cosmology. As a result, the notion of the sanctity of life and the immorality of violating the sacred is often dismissed as mere fantasy, subjectivity, or poetic expression, without a basis in objective knowledge. To remedy this, we must return to a worldview that recognizes the cosmic dimension of ethics and allows us to expand our ethical considerations beyond just human beings within our own community, state, society, or religion. This will enable us to approach ethical dilemmas with both our hearts and souls, extending our moral obligations to non-human beings and all human beings alike.[691]

According to Nasr, ethics and divine principles are closely intertwined. He asserts that in conventional communities, human laws were believed to be in harmony with universal laws, and emphasizes the need to reestablish this crucial connection between human and cosmic law.[692]

691 *Sophia*, 2004, p.10.

692 *Sophia*, 2004, p10-11.

The potential implications of the assertion of "divine law" raises important questions. How can humans understand and interpret this concept? How will it be implemented in human societies? Even within traditional societies, different interpretations and systems have been developed based on "divine" law. This can vary greatly between different cultures, such as re-establishing principles of Manu and its caste foundation in Hindu culture or implementing sharia law in Islamic cultures. The specific implementation is not discussed in detail in Nasr's publications, although it is addressed in his publications aimed at Islamic audiences, as we will see in the next chapters.

Secondly, traditional systems emphasize the significance of human development. Humans are seen as imperfect and have the potential to improve their character and soul, ultimately reaching a spiritual connection to the divine. Many traditional religious systems provide guidelines for this personal growth, including an ethical and spiritual shape of the "Perfect Man" and training to achieve this ideal. The methods provided allow individuals to control their impulses and acquire virtues such as contentment, discipline, and self-control.

To reach spiritual enlightenment and live in harmony with nature, virtues such as contentment are essential for mankind. Contentment in particular plays a crucial role in curbing greed, the constant desire to acquire more possessions. As pointed out by Nasr, this drive is a fundamental aspect of consumerism that fuels the economic system. "These traditional virtues which have allowed countless generations to live in equilibrium with the world around them were at the same time conceived as ways of perfecting the soul, as steps to the perfection of human existence."[693] As a result, improving a person's inner condition will also have an impact on how that person interacts with the outside world: inner harmony reflects that of the exterior world.

A disruption in the inner state of individuals can also have negative effects on the external world, as seen in environmental issues. This connection is referred to by Nasr as the misdirection of the soul's yearning.[694] The human

---

[693] Nasr, S.H., "Religion and the Environmental Crisis", in: Chittick (eds), "*The Essential Seyyed Hossein Nasr*", p. 33.

[694] Ibid., p.272.

desire to seek the divine and the infinite, often repressed in modern secular perspectives, continues to drive individuals despite the lack of fulfillment, ultimately leading to the environmental crisis. When individuals focus on material pursuits, it can lead to advancements in science, technology, and economic development, but also cause harm to the environment. Instead of progress being directed towards spiritual fulfillment, it has become solely centered on material gain. This has left individuals with a sense of emptiness and searching for fulfillment in the material world. True freedom and fulfillment come from spiritual realization, and efforts should be directed towards fulfilling inner spiritual needs rather than solely material pursuits.

> .... The insatiable thirst of man leads him to constantly seek fulfillment in the material world, yet he is never content with what he possesses in the physical realm. In his pursuit, he continually exerts his boundless energy upon nature, disrupting its natural order and transforming it into the chaotic and unsightly state we observe today in numerous regions of the world, all of which bear the unmistakable imprint of modern human activity. Spiritual creativity has been replaced by a relentless pursuit of technological advancement, resulting in the creation of countless gadgets and products that leave behind a trail of waste and desolation, perpetually expanding.....[695]

Sacred Nature

The traditional worldview also holds that nature has a sacred quality, which differs from the modernist perspective of nature as a machine, a collection of quantities in motion, or simply a source of raw materials.

> The sanctity and vertical aspect of nature are acknowledged by all conventional religions. Nature is regarded as possessing spiritual attributes or considered sacred. Regardless of their culture, mystics have recognized nature as an embodiment of the divine without being ensnared by pantheism. Certain mystics have even referred to the earth as an angel or Mother. For centuries, people have known that nature is not a lifeless collection of elements, but

[695] Ibid., p.272.

> a living entity with its own rhythms and harmonies. This concept, known as the Gaia theory, is an ancient idea that is currently being revisited.[696]

The presence of divinity is reflected in nature's creative manifestation, as evidenced by its various phenomena. Nature's beauty can be perceived as a theophany, an expression of divine order. Those who possess wisdom can discern spiritual significance in the natural world, interpreting a forest's trees as symbols of paradise and a lofty mountain peak as a representation of the sacred mountain situated at the universe's center.[697] According to Nasr, modern society has lost its connection to the spiritual realm and is unable to see the deeper meanings in nature. Nature is no longer viewed as sacred and holds little spiritual significance. This is seen as a major loss as nature can serve as a teacher for humanity, providing lessons in morality, intellect, and spirituality in addition to the more tangible knowledge it can provide.

> The natural order speaks to humanity's fundamental needs and ultimate purpose, even if this purpose goes beyond the external aspects of nature. The various religious teachings that expound on this order serve to enrich the message that humanity receives, reminding us of forgotten truths within a specific cultural context. Although we are ultimately destined for the invisible realm of the spirit, there is still much to gain from observing the natural order. Some Sufis have proposed that the universe can act as a mentor to aid us in surpassing the constraints of the universe itself.[698]

The universe can be likened to a book where every event holds significance both within and beyond its physical form. The language utilized in this natural book is symbolic, and to comprehend its secrets, we must reacquire knowledge of the symbolism of nature. Unfortunately, we have lost the ability to understand this language and the message it conveys. Thus, it is vital to acknowledge the innate sacredness of nature, not merely by declaring it, but through a realization of its inherent sanctity. This can be achieved through a process of rediscovery or recollection of the sacredness of nature.

---

696 *Sophia* 2004, p.11

697 Nasr, S.H. "*Religion and the Order of Nature,*" 1996, p.285.

698 Ibid., p.65.

> It is essential to rediscover our perception of nature as a fascinating truth to which we are intricately connected, not only on a physical, chemical, and biological level but also on a psychological, intellectual, and spiritual level. To achieve this, we must expand our understanding of nature beyond the limitations of quantitative science while still acknowledging the valuable discoveries made in the quantitative and material domains.[699]

An inner transformation is crucial for rediscovering the enchantment and sacredness of nature. With an understanding of the religious order of nature, one can regain their inner sense of the sacred. Spiritual training, personal development, and spiritual realization are also essential in this process.

> Those who perceive the divine light within themselves also witness its reflection in the natural world. However, the process of resacralizing nature is contingent upon our recognition of our identity and purpose within the world. It requires a collective awakening by humanity to comprehend who we are and what we are doing in this world.[700]

Nasr suggests that the sacredness of nature is best conveyed through narrative:

> Several years back, we had the privilege of strolling alongside 'Allamah Tabatabai, a distinguished philosopher, spiritual leader, and revered saint hailing from Persia, in a scenic valley situated in the Alborz Mountains just outside of Tehran. After completing our morning prayers, we felt a profound spiritual aura in the tranquil natural surroundings. During our walk, the master remarked that if even a single or a couple of "uninitiated" individuals without prayer and an affinity for nature were to appear, the entire atmosphere would alter, and the spiritual essence of nature would vanish. Within a few minutes, his words were validated when two city dwellers with a modern outlook and demeanor emerged from around the bend in the river. Suddenly, the ambiance shifted,

---

[699] *Sophia*, 2004, p.12.

[700] Nasr, S.H. "*Religion and the Order of Nature*," 1996, p.286.

> and something was obscured. It was akin to a stranger abruptly entering a conventional Muslim household, prompting the women to hastily veil themselves to conceal their beauty from the unfamiliar gaze. The master grinned and expounded that this is precisely what transpires when outsiders penetrate nature's sanctuary; nature hides her most intimate beauty from them.[701]

Nasr suggests that there is a spiritual dimension of reality that can be perceived by those who have been trained to do so. However, those who are not spiritually trained, or "profane," are unable to see this aspect of the world. He also states that nature has an active role in this and that it "responds" to the spiritual state of humanity by "hiding" or becoming obscured. According to Nasr, the relationship between humanity and nature is reciprocal. If humanity becomes increasingly disconnected from the spiritual aspect of nature, then nature will suffer. Nasr believes that the environmental crisis we face today is a consequence of modern man's loss of connection with nature's soul.

Man occupies a crucial position in the hierarchy of beings, and nature relies on his existence. As per Nasr's philosophy, humanity is a conduit of blessings for nature, and its existence on this planet enables a spiritual linkage between nature and the divine sphere. While modern man is not indispensable for the world's survival, mankind's unchanging and timeless essence is crucial for its continued existence.[702]

> "Human beings are conduits of grace and light within the natural world. As a result, their responsibility is weighty," the author states. Humans have been bestowed with the ability to govern nature, but also have the capacity to devastate it. Consequently, their actions hold significant ramifications in the cosmos.[703]

The link between a person's inner state and the state of nature is not immediately obvious. It is easy to understand that reducing consumerism by promoting virtues such as contentment, discipline, and self-control is a kind of virtue ethics. Furthermore, Nasr argues that this connection goes

[701] Ibid., p.292, n30.

[702] Spanish interview Agenda Viva, in: Sophia, 2006.

[703] Nasr, S.H. "*Religion and the Order of Nature,*" 1996,, p278.

much deeper, stating that without man nature cannot exist. This may seem implausible from a modern perspective, but Nasr claims that there are "esoteric, metaphysical and universal reasons" for this relationship.[704] However, he does not provide any further explanation in the Spanish interview.

According to Nasr, the relationship between humans and nature is deeply interconnected. He argues that when we only view nature as physical, we disconnect it from its spiritual and metaphysical aspects, causing a loss of balance.[705] To restore this balance, Nasr suggests that religious rituals can be used to reconnect the earth with higher levels of reality, thereby re-establishing harmony with the cosmic order. According to Nasr, certain religious rituals hold the power to reconnect humanity with nature and the divine. Examples of these include the Eucharist in Christianity, the five daily prayers in Islam, and traditional rites in Hinduism, Buddhism, and Native American religions. According to Nasr, a community cannot coexist in coherence with the environment, the greater celestial orders, and ultimately, with God, unless it establishes a ritualistic connection with the natural world.[706] Nasr acknowledges that his ideas may be difficult for modern audiences to grasp and therefore avoids discussing them in certain settings, such as the San Rossore Conference on Climate Change.[707] However, in his works *"Religion and the Order of Nature"*[708] and other writings intended for traditionalist readers, he places a strong emphasis on the ritual aspect of his ideas.

Nasr believes that the main reason for the environmental problem is a flawed modernist perspective and offers an alternative worldview rooted in traditional religious metaphysics reinterpreted for contemporary times. He argues that a deeper understanding of these principles can lead to a restoration of harmony in humanity and the environment. Nasr offers

---

704 Interview *Agenda Viva*, in *Sophia*, Vol 12, No2, 2006, p29.

705 Nasr, S.H., "Religion and the Environmental Crisis", in: Chittick (eds), *"The Essential Seyyed Hossein Nasr,"* 2007, p36.

706 Ibid., p.36-8

707 Nasr, S.H., "Religion and the Environmental Crisis", in: Chittick (eds), *"The Essential Seyyed Hossein Nasr,"* 2007, p.34-36.

708 Nasr, S.H. *"Religion and the Order of Nature,"* 1996, p278-282. Nasr, S.H., "Religion and the Environmental Crisis", in: Chittick (eds), *"The Essential Seyyed Hossein Nasr,"* 2007, p.34-36.

several solutions to this problem.

(1) The Key to Traditional Knowledge

(2) The Key to Religious Ecological Ethics

(3) The Key to the training of Spirituality and Ritual

Religion can be considered the master key as it holds answers to many questions. Its theological, practical, and ceremonial dimensions play a critical and fundamental role. Understanding the principles of true reality is crucial to finding solutions as ignorance can lead to misguided actions and chaos. Major religious traditions hold this understanding at their core. To fully comprehend the relationship between religion and the natural order on a universal scale, it is essential to explore multiple traditions, especially given the current global threat to the natural order. Nasr's book, Religion and the Order of Nature aims to uncover the fundamental principles that govern reality, not only in a scholarly sense but also in a practical sense. Nasr hopes to provide insight into the underlying causes of the present ecological problem and ultimately find a solution at its root level.[709]

Nasr believes that strong environmental ethics must have a religious foundation, referred to as a "moral order." He argues that without this grounding, ethics lack a stable reference point. Therefore, he believes it is essential to incorporate religious principles into environmental ethics in order to establish a sound foundation. According to Nasr, the predominant skepticism towards religious ethics is a significant barrier to resolving the environmental crisis. He believes that many Western intellectuals approach environmental issues as if all individuals were non-religious and adhere to secular philosophy. They then attempt to create a logical environmental ethics, despite its likely ineffectiveness in addressing the environmental crisis.[710]

Nasr highlights the challenge of creating environmental ethics that can match the widespread appeal of religious ethics, given the predominance of religious beliefs among the world's population. To address this,

---

[709] *"Religion and the Order of Nature,"* 1996, p.25, n2

[710] Ibid., p30.

he proposes an environmental ethic that is grounded in divine qualities like satisfaction, obedience, and self-control and centered on God (virtue ethics). Such an approach would help to counter consumerism, reform economic structures, and promote environmental harmony. Moreover, it would facilitate a shift in humanity's focus from material possessions to the divine, redirecting the soul's potent energies towards a higher purpose.[711]

In addition to providing knowledge and ethical guidelines, religions also offer techniques and practices to achieve spiritual quality and reach a form of inner perfection.

> For thousands of years, religions have grappled with the challenge of conquering the passionate ego, often symbolized by the inner dragon. One such symbol is St. Michael's slaying of the dragon with his lance, which conveys the idea that only the lance of the Spirit can defeat this dragon, also known as nafs in Sufism, or the lower soul within us. Today, this issue is seldom considered, and we must ask where St. Michael and his lance are. How can we prevent people from constantly craving more and more without relying on the power of the Spirit that religion provides? Rational arguments alone cannot persuade people to use less, become less covetous or be less greedy. Only religion has the ability to achieve this without resorting to physical coercion.[712]

Religious traditions provide time-tested techniques and the collective knowledge of generations, enabling individuals to control, mold, and regulate their internal impulses, with avarice being the most enduring challenge. The inner self, or the "nafs" in Sufi teachings, which encompasses the passionate ego and the base instincts within us, can only be overcome by religion. Moreover, spiritual growth is essential for developing the skills required to rediscover the divine.

How can the ideas presented by the traditional paradigm be put into practice to address the current crisis? What actions are necessary for modern humanity to take in order to solve the crisis, and what is the agenda proposed

---

[711] *Sophia*, 2004, p10.

[712] Nasr, S.H., "Religion and the Environmental Crisis", in: Chittick (eds), "*The Essential Seyyed Hossein Nasr*", 2007, p33.

by Nasr? Who bears the responsibility of implementing measures, where and when, and what regulations and institutions are necessary to tackle the ramifications of environmental ethics on environmental policy, climate change, pollution mitigation, energy preservation, species preservation, and the broader structuring of society, encompassing economic and financial systems?

Nasr firmly believes that reinstating the essential principles, and above all, adhering to them, will help humanity restore a healthy equilibrium with the environment. This belief is founded on two fundamental assumptions: (1) that traditional cultures coexisted harmoniously with their surroundings, and (2) that the organization of these societies is a testament to their traditional perspectives and fundamental values. In fact, it is widely held that the principal drivers of these civilizations were their respective religions.

> The source of all the ancient religions that have shaped human existence throughout history, giving rise to traditional societies with their sacred customs, social structures, and artistic and scientific achievements, is a single divine reality.[713]

Nasr seems to assume that re-establishing those values and reorganizing our modern society in accordance, will lead to a revitalized communal lifestyle that is in peace and balance with the environment. It is sad that Nasr doesn't go into more detail on this subject in his articles given how important it is to the argument. It is essentially portrayed as an obvious reality. It can be faulted in a number of ways. For instance, were traditional communities better able to coexist with their natural surroundings? *Collapse* (2006)[714], a new book by Jared Diamond that explores the factors that led to the demise of early civilizations, substantially casts doubt on this claim. This assertion has already been disputed by others. Second, is there a direct relationship between social structure and the prevailing ethics and worldview?

The disconnect between beliefs and actions, as well as the relationship between ethics and actions, has been a topic of discussion among environmental philosophers and others. Additionally, the question of whether

---

[713] *Sophia.*, 2004, p.12

[714] Diamond, J., "*Collapse*," New York, 2006.

societies can be defined by their core principles has also been debated. This perspective suggests that societies can be seen as distinct entities with certain inherent qualities.[715] The subject has sparked scholarly discussions among sociologists, historians, and comparative religion experts, and has also been a topic of contention surrounding the idea of Orientalism. Although we may delve into these topics in the future, for the time being, we will primarily need to disregard them.

Emphasizing the potential political and social consequences, it should be noted that Nasr's proposal for a substantial overhaul of society, rooted in traditional values, holds great significance. This infers to the query of how these principles can be applied in a post-modern society, which has been previously discussed in relation to ethics and divine law. However, some critics contend that Nasr's vision involves reverting to a medieval feudal system.[716] We will examine Nasr's understanding of Islam and address these criticisms further in that discussion.

Nasr's main emphasis lies in comprehending the intellectual components of the environmental crisis, with a particular focus on the metaphysical and cosmological elements. Although he recognizes the necessity for practical solutions, he delegates the responsibility of devising them to others. Despite this, during several interviews, he subtly expresses a predilection for traditional, small-scale technologies, farming practices, and eco-village communities, which are also advocated by certain Western environmentalists. In an interview with Muzaffar Iqbal, Nasr cites these groups as an example:

> A movement is underway to raise awareness among farmers, builders, and the general public, with a focus on promoting sustainable practices. Two examples of this effort include the Resurgence Journal in England and the Schumacher School in Devon, England. As mentioned earlier, the construction of environmentally sustainable villages that preserve traditional agriculture and architecture

---

[715] For the distinction between ethics and action see "Calicott" J. Baird', in: Taylor, B. (eds), "*Encyclopedia of Religion and Nature,*" 2 volumes, New York: Continuum, 2005, 2008. p.252-253.

[716] Upton, Charles, "What is a Traditionalist? Some Clarifications", accessed *Sacred Web*:www.sacredweb.com. This is a response to criticism of the Traditionalist perspective by Dr. Hajj Muhammad Legenhausen 'Why I am not a Traditionalist', accessed at: http://religioscope.com/info/doc/esotrad/legenehausen.htm.

> requires significantly less energy than modern towns and villages. Therefore, it is imperative for us in the Muslim world to observe and learn from the progress being made in the West, as well as in India and China to some extent.[717]

Nasr emphasizes that a broader perspective is necessary for effective environmental legislation. In his view, a comprehensive understanding of nature is crucial for such laws. He also contends that societal transformations are primarily driven by worldviews, which have the power to shape social and economic conditions. According to Nasr, it is often a select group of influential individuals who drive these changes. He cites the seventeenth century, during which a few people, including Galileo, Newton, and Kepler, proposed a new physics that transformed the worldview of Western civilization.[718]

> Back then, there was no awareness that the Western worldview was founded on a mechanistic view of nature, which ultimately resulted in its destruction. This idea was not considered by anyone in the eighteenth or nineteenth century, but forty years ago, I was one of the first to predict it.
>
> In my opinion, the process of a select few influencing the majority is a common occurrence. The influence of Plato and Platonism, as well as the teachings of Christ, the Prophet of Islam, and other great religions, have persisted throughout history and continue to impact society today.[719]

The primary means by which a worldview is put into action is through the actions of a select group influencing the actions of the larger population. A spiritually enlightened group will lead the way in bringing about the necessary changes to protect our world. These individuals will serve as inspiration for others to embark on their own journey towards spiritual and personal growth. It is often said that one must first change themselves before one can make a meaningful impact on the world.

---

717 "The Islamic perspective on the environmental crisis: Seyyed Hossein Nasr in conversation with Muzaffar Iqbal."

718 "Transcript of the interview with Seyyed Hossein Nasr on 31 October 2008."

719 Ibid.

> In the pursuit of reforming the world, many individuals make the mistake of neglecting their own personal development. Marxism and other similar ideologies have fallen victim to this error in recent history. However, when it comes to environmental issues, it is crucial to begin by working on oneself. It is essential to cultivate a sense of respect for all living creatures, acknowledging their inherent rights and value.[720]
>
> As stated in the Qur'an, animals and creatures of the earth are nations like us, and the term Umma in Islamic thought applies to them as well. It's essential to acknowledge that the natural world is infused with significance, essence, and sagacity, and as humans, we are not its masters but rather its guardians. In order to assert dominion over nature, we need to prioritize upholding the rights of all living beings. This is a cornerstone of my philosophy.[721]

Nasr differentiates between the elite and the general population. He believes that only a select group can fully embrace the spiritual path, however, the benefits of this path extend to society as a whole, impacting culture and societies.

> The number of individuals in Japan who have experienced satori, or Buddhist illumination, is relatively small. However, the impact of Zen on Japanese culture and its views on nature, gardening, and other aspects of life extends beyond the elite. It permeates Japanese culture as a whole. Similarly, in Islam, while there have been great Sufis such as Jallauddin Rumi, their spiritual messages have been disseminated through literature to the broader society, much like the teachings of great Christian saints.
>
> As a Persian myself, I know that Sufism is deeply ingrained in Persian literature. While only a select few may be called to fulfill a spiritual role, if they carry out their function effectively, their influence can radiate throughout society as a whole, as Christ stated when he said, "Many are called, but few are chosen."[722]

[720] Ibid.

[721] Ibid.

[722] Ibid.

Chapter Summary

Religion plays a crucial role in transforming the perspectives and lifestyles of the general population. As a majority of individuals around the world still hold religious beliefs, Nasr argues that the most widely accepted form of ethics is "religious ethics." Without the influence of religion, what other means can be used to deter the consumerist trend among the masses? Religion, specifically the concept of the "fear of God" and the belief in punishment in the afterlife, is the only force that can effectively motivate the masses to take action. In the Islamic world, traditional religious scholars, or *ulama,* play a crucial role in promoting this type of thinking. To fully understand the implications of these ideas, it is necessary to examine the specific views of the author, Nasr, on Islam and Islamic environmentalism.

What about the Western population's nonreligious majority? Western civilization is alive with a variety of organizations and movements that combine spirituality with environmental goals. As examples of sustainable living, several of them also include "traditional or indigenous" cultures. We'll find that Nasr has criticism for these kinds of protests. These practices that pass for "science" are actually "New Age pseudo-science of the universe."[723]

[723] Nasr, S.H., 'Religion and the Environmental Crisis, in: Chittick (eds), *The Essential Seyyed Hossein Nasr*, p.38-39.

Conclusion

Nasr's work proposes that the dominant modern scientific perspective, grounded in "rationalism and empiricism", is primarily responsible for the rise of "scientism and scientific progressivism," ultimately leading to the present-day ecological problem. From an epistemological perspective, Nasr contends that the dismissal of revelation as a means of acquiring objective knowledge about the natural world and the negation of Intellect as a superior form of perception, which are inherent in "scientific rationality and rationalism", give rise to "scientific progressivism." The transition from the pre-modern era to the current worldview has neglected three essential metaphysical principles prevalent in various world religions: the oneness of existence, the gradation of existence, and the spiritual teleology of nature. According to Nasr, this shift has resulted in a spiritual impoverishment of human consciousness and has had catastrophic consequences for the environment.

Nasr posits that the answer to the environmental crisis lies in the re-establishment of a worldview that incorporates perennial metaphysical concepts. Therefore, the solution is not based on calculating the pros and cons of various external methods to decrease environmental destruction. Its goal is to encourage people to shift their perspective of nature beyond simply a provider of resources or a source of pleasure, and instead recognize it as a sacred entity with an inherent connection to humanity. According to Nasr, the way in which we view nature is critical, as our thoughts and actions are inevitably influenced by our understanding of reality, even if it is only loosely defined or unconsciously held.

Esotericism in Islamic Tradition with a Traditionalist Perspective

Nasr is advocating for an environmental perspective in Traditionalist Islamic thought, which draws from Esotericism. To understand his proposed solution to the current ecological crisis, it is crucial to define and clarify the terms 'Traditionalist', 'Islamic', and 'Esotericism'. His advocacy centers on a return to orthodox religious beliefs, contrasting with the secular worldview prevalent in the Western world. This book aimed to expound Nasr's eco-philosophy as a viable solution to the unsustainable state brought on by modern industrialization. Nasr posits that the ecological crisis is rooted in a deeper spiritual crisis, where modern humanity has severed its connection with the divine and spiritual aspects of both humanity and nature. Through the denial of the divine center of existence and the interconnected hierarchy of realms, modern humanity has been left aimless in an unmeaningful universe. Modernity, with its rationalistic, reductionist, materialistic, and anthropocentric worldview, drives cultural, economic, and technological progress but only offers a limited, flawed understanding of reality through its emphasis on "absolute" knowledge.

Nasr argues that the contemporary paradigm fails to recognize the elevated levels of reality within the Great Chain of Being, leading us down a misguided path that ultimately leads to self-destruction. This flawed worldview is the root cause of the ecological crisis, with the economy, driven by institutionalized greed and consumerism, being the immediate cause that reduces humans to mere materialistic beings. To prevent catastrophe, we must make radical changes and shift our perspective, as Nasr believes that this can only occur through a rediscovery of the traditional worldview and reestablishing balance and harmony with nature. This requires a change in our views of the world, ourselves, and our relationship with nature. Nasr posits that in the industrial age, Western culture lost the traditional understanding of the true nature of existence, which was grounded in religious beliefs. He contends that all genuine religions, including but not limited to Buddhism, Hinduism, Taoism, Confucianism, Judaism, Christianity, Islam, and the Indian American traditions, share a collective set of common metaphysical ideologies that reflect an eternal and absolute truth. These principles facilitate a harmonious coexistence with the hierarchy of existence and nature.

Nasr conducted a comparative examination of nature theologies from various religions and identified five core principles that are present in all authentic traditions. These principles include: (1) recognizing the existence of a solitary and absolute divine truth, (2) acknowledging the hierarchical organization of the universe, (3) acknowledging the presence of cosmic or divine laws governing all aspects of reality, (4) recognizing the inherent interconnectedness of all existence, and (5) acknowledging the importance of humanity within the Great Chain of Being. Nasr recommends that reaffirming these principles can help restore the balance between human society and the natural world.

This study proposed that Perennialism, Traditionalism, or Esotericism, as defined by prominent figures of the Traditionalist school including Ananda Coomaraswamy, Frithjof Schuon, and Réné Guénon, represents a particular kind of thought. This traditional worldview shapes our understanding of the world and ourselves, and in turn, influences our choices and actions. To live in harmony with the environment, ethical principles must align with divine or cosmic laws, which regulate all planes of presence, including the physical, spiritual, religious, and divine. Human actions and freedom are naturally limited and controlled by this prevailing order. Practical ethics must acknowledge this order, and embrace values such as self-restraint, self-control, balance, and satisfaction. Additionally, this worldview underscores human accountability regarding all living creatures in nature, as all beings occupy a rightful place in cosmic harmony. Religion provides guidance on ideal human behavior and the means to achieve self-development. Practicing virtue and cultivating the soul can counteract consumerist greed and encourage lifestyles that respect the natural limits of our planet. Nasr argues that religion offers a solution to the crisis by promoting theocentric virtue ethics based on the revealed consecrated word of the heavenly order of reality (metaphysics).

Environmental Ethics in Islamic

Although all religions may have a common universal core, we should not make the mistake of assuming that all religions are the same or that we can pick and choose from their diverse traditions without caution. Nasr, along with other traditionalists, stresses the importance of following a single path and adhering to orthodox religious practices. Nasr adheres to a traditional interpretation of Islam that is firmly rooted in the classical Sharia'ite injunctions derived from the Qur'an and Hadith. This interpretation does not involve a radical reevaluation of these sources based on modern theories, nor does it adopt a fundamentalist approach that seeks to return to an idealized version of Islam. Instead, traditional Islam embraces the rich cultural legacy of Islamic history, which spans theology, philosophy, literature, science, arts, and architecture. According to Nasr, Sufism embodies the innermost essence of Islam as it represents the esoteric aspect of the religion. The metaphysical doctrines of Sufism are genuine expressions of perennial wisdom or gnosis, which serves as the shared metaphysical foundation of all religions. Nasr holds the view that the teachings of Ibn Arabi best exemplify this wisdom and have significantly influenced subsequent Sufi thinking. Nasr's own ideas are shaped by a fusion of Persian Shi'ite mysticism and Ibn Arabi's metaphysical principles.

Nasr's Islamic environmental ethics and teachings are deeply entrenched in his understanding of traditional Islam. These teachings encompass the Quranic and Hadith teachings on the interplay between humans and nature, environmental laws derived from sharia, traditional social, political, and economic institutions, as well as spiritual rituals aimed at harmonizing the soul with the divine order and re-establishing equilibrium with nature.

Criticisms of Nasr's Idea of Islamic Ecological Ethics

Nasr and other proponents of Islamic environmentalism have encountered opposition to their belief that Islam is intrinsically environmentally conscious, and that Muslims simply need to acknowledge this facet of their religion. Two prominent critics of this perspective are Richard Foltz and Kaveh Afrasiabi, with Afrasiabi adopting a more aggressive tone than Foltz. Despite Afrasiabi's more forceful objections, my response to Foltz's more nuanced critique should address both critics' concerns.[724] Due to the scope of this study, only a few of the numerous, unconvincing points that Foltz cites to argue against the notion that Islam has always been eco-friendly as claimed by Nasr and other Islamic environmentalists will be mentioned. Nasr clarifies that the Qur'anic verse "In Whose Hand is the dominion (*malakut*) of all things" (23:88) connotes not just the administration of all things by God, but also the presence of the essence of all things in God's Hand.[725] However, Foltz argues that Nasr's definition of nature differs from the way in which Ibn Sina or Ikhwan al-Safa used the term, and this discrep-

[724] See Richard C. Foltz, "Islamic Environmentalism: A Matter of Interpretation," in "*Islam and Ecology: A Bestowed Trust*", eds. Richard C. Foltz, Frederick M. Denny, and Azizan Baharuddin (Cambridge, MA: Harvard University Press, 2003), 249-79

[725] Seyyed Hossein Nasr, "*Religion and the Order of Nature*" (New York and Oxford: Oxford University Press, 1996), 61.

ancy is evidence of the varying perspectives of Islamic environmentalists.[726] Additionally, the concepts of unchangeable prototypes (*al-a'yan al-thabita*), which Ibn 'Arabi's understanding of several Qur'anic passages credits, and the notion of a hierarchy of reality are directly related to Nasr's statement.[727]

Foltz rejects the view of Islamic environmentalists who interpret the Qur'anic concept of humans as the *khalifah* (vicegerent) of God to imply that humans are tasked with preserving nature in line with God's wishes, acting as caretakers executing His plan on earth. Foltz doesn't provide any substantiating evidence for his position.[728] Abu Ja'far al-Tabari (d. 923) and Abu 'Abdallah al-Qurtubi (d. 1273), who were traditional Qur'anic interpreters, interpreted the term *khalifah* as a person responsible for implementing God's will. Likewise, Sufi scholars, including Ibn 'Arabi and Nizam al-Din Nishaburi (d. 1327), believed that the *khalifah* had the ability to evaluate God's creation using His attributes.[729] [730]

Despite the immense influence of modern Muslim environmentalists, ancient Qur'anic commentators, and mystics like Ibn 'Arabi in interpreting Islamic sacred scriptures, Foltz rejects their explanations as unauthentically Islamic. He claimed that non-hierarchical explanations of the sacred scriptures are adequate for today and that only accepting one explanation as true while excluding others is contradictory to his belief.[731] Foltz's stance on the definition of "*khalifah*" terminology is puzzling because Islamic environmentalists commonly agree on its meaning. While others define it in a non-hierarchical sense, Foltz believes that it implies a problematic hierarchical relationship between humans and the environment. Interestingly, this interpretation aligns with the views of Muslim reformers and some fundamentalists who wrote in this manner under the influence of European Humanism and in response to allegations that Islam doesn't value

726 See Nasr, *"An Introduction to Islamic Cosmological Doctrines"*, 8.

727 Mahmoud M. Ayoub, *"The Qur'an and Its Interpreters,"* Vol. I (Albany, NY: State University of New York, 1984), 78-79.

728 Foltz, "Islamic Environmentalism: A Matter of Interpretation," 253.

729 See Ayoub, *"The Quran and Its Interpreters,"* 74 and 76.

730 Ibid., 78-79.

731 Foltz, "Islamic Environmentalism: A Matter of Interpretation," 249.

human beings enough, as noted by Othman Llewellyn.[732] [733] In addition, in the context of what we've just discussed, the only condition in which one has any control over the established order is one of submission to God's will. In this interaction between humans and nature, exploitation of the natural world is not acceptable.

According to Foltz, Islam, like Christianity and Judaism, is anthropocentric, which is reflected in his dismissal of the environmentalist interpretation of "*khalifah*." He believes that the Islamic faith prioritizes the relationship between humans and God over other aspects, such as the significance of non-human creatures. Foltz's oversight is that the development of virtues such as justice, love, kindness, contentment, and simplicity not only strengthens the connection between humanity and God but also safeguards the environment from human exploitation, as highlighted by environmentalists Sachs, Speth, Meadows, and others. Additionally, the Qur'an and hadith emphasize the importance of the natural world and the obligation to display compassion for all living creatures.[734] [735] Additionally, traditional societies did not require a more in-depth theology about the environment without industrialization. According to Ali Ahmad's observation, Islam's influence in Northern Nigeria is acknowledged to have contributed significantly to raising awareness about the natural world, including its beauty, intricate organization, and interdependence. This suggests that other traditions did not necessarily conserve their natural worlds better in the pre-modern era than Islam did. Moreover, there is scant proof to indicate that Hindu India or Buddhist Sri Lanka are any less complicit in the degradation of their present-day environment than their neighboring Muslim countries, Pakistan and Bangladesh, with whom they share physical and cultural boundaries.[736] [737] [738]

---

[732] Ibid., 254.

[733] Llewellyn, "The Basis for a discipline of Islamic Environmental Law," 190. For more, see Chapter 4.

[734] Foltz, "Islamic Environmentalism: A Matter of Interpretation," 257.

[735] Othman Llewellyn, "The Basis for a discipline of Islamic Environmental Law." 249-279.

[736] Ali Ahamd, "Nigeria," 76.

[737] Foltz, "Islamic Environmentalism: A Matter of Interpretation," 255.

[738] C02 emissions per capita based on 2007 data for India, Sri Lanka, Pakistan and Bangladesh were 1.18, .64, .85 and .25 tonnes respectively. See "*C02 Emission From Fuel Combustion* report of International Energy Agency(IEA)," 2009 edition,

Foltz contends that the historical understanding of the Qur'anic interpretation of tawhid did not encompass all aspects of reality, despite modern Islamic environmentalists like Nasr arguing otherwise. Nasr's argument against the fragmented view of reality presented by modern science relies on tawhid as the comprehensive unity of reality.[739] However, Foltz's arguments require clarification. According to Foltz, the only way to fully comprehend tawhid as an all-encompassing reality is by utilizing the concept of 'unity of Being' (*wahdat al-wujud*), which is linked to the teachings of Ibn 'Arabi. He believes that this notion is founded upon a single Qur'anic verse, "Wherever you turn, there is God's countenance," and that various Sufis have had comparable experiences even prior to Ibn 'Arabi. However, Foltz's claim that Ibn 'Arabi subscribed to monism is contradicted by the doctrine of the stable archetype (*al-a'yan al-thabita*) that is associated with 'unity of Being'.' [740] [741] Furthermore, evidence suggests that the renowned theologian and Sufi, al-Ghazzali, held very similar beliefs to Ibn 'Arabi regarding *tawhid*. Al-Ghazzali makes it clear that the gnostics perceive nothing in existence except for the One, the Real, and are fully absorbed in the absolute unity where only God exists. This assertion by al-Ghazzali lends credence to Ibn 'Arabi's elucidation of tawhid and places him as a noteworthy predecessor in this regard.[742] [743] [744]

Although Foltz recognizes the considerable impact that Ibn Arabi had on Sufism, he attempts to discredit his metaphysics as non-conformist and contentious. Nevertheless, Foltz appears to be oblivious to the fact that Sufism was indistinguishable from traditional Islam for the majority of

---

p. 89.

739 Foltz, "Islamic Environmentalism: A Matter of Interpretation," 253.

740 Qur'an 2:115 "*The Message of the Quran*", trans. Muhammad Asad, Bilingual edition (Watsonville, CA: The Book Foundation, 2003).

741 Reza Shah-Kazemi, "*The Other in the Light of the One: The Universality of the Qur'an and Interfaith Dialogue*" (Cambridge, UK: The Islamic Texts Society, 2006), 77. For Sufi experiences of unity of reality, see "*Knowledge of God in Classical Sufism: Foundations of Islamic Mystical Theology*," trans. John Renard (New York: Paulist Press, 2004).

742 Foltz, "Islamic Environmentalism: A Matter of Interpretation," 250.

743 Foltz, "Islamic Environmentalism: A Matter of Interpretation," 256.

744 Abu Hamid Muhammad al-Ghazzali , "*The Niche of Lights*, trans. David Buchman" (Provo, Utah: Brigham Young University Press, 1998), 17. See al-Ghazzali, "*The Niche of Lights*," 20.

Muslims until the mid-19th century. William Chittick, a leading Western advocate of Ibn Arabi's concepts, clarifies why Ibn Arabi has been the most noteworthy figure in Islamic philosophy following al-Ghazzali, and how his influence experienced a sharp decline during the latter half of the 19th century:[745] [746] [747]

> Following his passing in 1240, the teachings of Ibn 'Arabi rapidly disseminated throughout the Islamic world. This dissemination was not due to the mystic nature of his teachings, which impressed the Muslim elite shaped by various forms of rational discourse. On the contrary, it was his well-supported arguments and extensive knowledge that convinced them to pay attention to him. Many of these scholars incorporated his fundamental perspectives and terminology, while others criticized or condemned his teachings. Regardless, Ibn 'Arabi was an influential figure that could not be ignored. His doctrines and viewpoints did not cater to a limited or exclusive audience but rather influenced all aspects of Islamic culture. Although Sufis may not have commonly read his works, individuals with scholarly interests, who often served as mentors and instructors, used a vocabulary heavily influenced by him and his close adherents. Due to his extensive recognition and far-reaching influence, Ibn 'Arabi became the subject of scrutiny by modernists and reformers in the latter half of the 19th century.[748]

James Morris, a renowned scholar in the field of Ibn 'Arabi and Islamic philosophy, notes that following Ibn 'Arabi's influence (at least up until the 18th century and the emergence of the modern West), the history of Islamic thought can be seen as a collection of notes and annotations to his work, as famously observed by Whitehead regarding Plato. Ibn 'Arabi's teachings also emphasize the critical role of adhering to the Shari'a for achieving spiritual fulfillment as a Muslim, although this perspective has been largely overlooked since the 19th century.[749] In a similar vein, Nasr contends that

---

[745] See Foltz, "Islamic Environmentalism: A Matter of Interpretation," 250-251.

[746] Ibid., 253.

[747] William C. Chittick, *"Ibn 'Arabi: Heir to the Prophets"* (Oxford: Oneworld, 2005), 1.

[748] Chittick, *"Heir to the Prophets,"* 2-3.

[749] James Morris, "Ibn Arabi and his Interpreters," Part II-A, *Journal of the American Oriental Society*, 106 (1986), 733.

Ibn 'Arabi's superior reasoning has played a significant role in the preservation of the Islamic tradition, as his esoteric teachings provide the necessary doctrines to safeguard the tradition against incorrect reasoning that can mislead individuals. For most people, intellectual intuition is not sufficiently powerful to counteract other human tendencies and prevent errors of the mind, making Ibn 'Arabi's contributions crucial for the continuity of the Islamic tradition.[750]

How to apply metaphysics in practice?

Nasr is focused on the overarching perspective or "grand vision" behind social systems. He believes that a shift in the collective mindset will lead to a transformation of social structures. He envisions a group of spiritually enlightened individuals taking the lead in promoting change and influencing others to adopt more sustainable lifestyles. Religious leaders, regardless of their own level of enlightenment, also have a crucial role in guiding the masses towards these goals. Nasr considers himself a revivalist in the realm of religion. Nasr's vision lacks specificity in many areas and raises questions that remain unanswered. Despite this, he shows a clear preference for traditional, small-scale technologies and eco-village societies identical to those proposed by figures such as E.F. Schumacher and Satish Kumar. He appears to align with environmentalist viewpoints in these areas. However, the exact social and political models he envisions as a result of his traditional paradigm are unclear.

Final Reflections

Nasr's solutions may appear idealistic, but they don't ignore the standard approaches of mainstream environmentalists. Instead, they go beyond those measures by outlining what needs to be understood and implemented. Despite Nasr's support for technology-based solutions and secular measures as a way to buy time for spiritual change, labeling his approach to the environmental crisis as unrealistic would be unfair.

[750] Nasr, "*Three Muslim Sages*" (New York: Caravan Books, 1964), 91.

Nasr is aware of the impacts of modern science and technology on nature and Islamic values through his criticisms of them, his promotion of traditional methods and production, and his vision of Islamic science as a replacement. By contemplating the Real (*al-haqq*) in every action and thought, he is truly being realistic. Hence, his approach is not impractical but rather firmly based on reality. In the Islamic ideology, the truth (*al-haqiqah*) must always prevail, and we ought to adopt a mindset of prioritizing truth over expediency, be it in political or other matters. It is essential to remember the verse from the Qur'an: "Say, 'Truth has come, and falsehood has departed. Indeed is falsehood, [by nature], ever bound to depart.'" (17:81)[751]

Nasr's belief that Sufism, metaphysics, and philosophy are crucial to protect against scientism may appear unconventional. Despite acknowledging the importance of certain areas, the author believes that only a small proportion of Muslims will excel in these domains. Instead, for the majority of Muslims, a fundamental comprehension and reverence for the spiritual dimension of their life and work would be adequate. Similar to the pre-modern era, having a select few individuals who excel in spirituality would act as role models and affirm the community's profound religious and metaphysical principles. Acknowledging the pivotal role played by these aspects of Islamic tradition, especially in modern times, is imperative. Providing opportunities for their study within society is equally important. This objective can be realized by incorporating Islamic philosophy and metaphysics into education.[752]

Fully embracing Sadra's natural philosophy for the implementation of Islamic science may take a significant amount of time and will necessitate spiritual training. It will focus as much on purifying the scientist's perception as on studying the natural world. The new Islamic technology will likely resemble traditional tools, as the emphasis is on keeping technology at a human scale. It may not have the external energy that defines modern science, but it will evoke greater spiritual dynamism towards realizing *tawhid*, the ultimate purpose of a Muslim's life. Likewise, the implementation of Islamic science is expected to encourage Muslims to move away

---

751 Seyyed Hossein Nasr, "Islamic Worldview and Modern Science," *Islamic Thought and Scientific Creativity*, Vol. 7 (1), 1996, p. 8.

752 Seyyed Hossein Nasr, *"The Need for a Sacred Science,"* (New York: State University of New York, 1993), 20.

from the pursuit of progress through material gain. Integrating Islamic science into contemporary practices can offer a spiritual orientation. This, in conjunction with a thorough awareness of the harmful effects of modern technology and the advantages of traditional tools, could result in a heightened appreciation for the latter in comparison to previous eras.

Nasr's Traditionalist position emphasizes the importance of preserving the authenticity of each religious tradition and promoting the development of a unique science based on its own natural philosophy. While the varying sciences of different religious traditions may appear to impede scientific collaboration in theory, Nasr's Islamic approach suggests that these sciences should adhere to timeless metaphysical principles and prioritize the spiritual growth of their practitioners. As a result, we can anticipate that cooperation among scientists from diverse religious traditions will occur in three ways: 1) at the plane of mystical awareness, 2) in terms of concepts, and 3) through technological advancements.

Spiritually aware scientists can share their understanding of the essence of various natural entities. Using this knowledge, combined with existing scientific discoveries, they can express their views on living in harmony with these entities. Sharing the secrets of simple, human-scale technologies should not pose a problem. If this level of understanding and cooperation is achieved, humans will not only save the environment but also their own souls.

Nasr is pessimistic about the immediate acceptance of his proposals by either the Muslim community or humanity in general. He believes that widespread apathy towards the environmental crisis may require major environmental catastrophes affecting modernized societies to prompt humanity to reconsider its attitudes and behaviors. Nonetheless, Nasr points out that only four centuries ago, Francis Bacon similarly doubted the adoption of his scientific inquiry method by the public.[753] Thus, what seems impossible in Nasr's approach now may become common in the future. As a believer in a God who can accomplish anything, Nasr retains hope.[754]

---

753 Francis Bacon, "*The New Organon,*" Aphorisms Book One, no. XXXIV in "*The New Organon and Related Writings*", ed. Fulton H. Anderson (New Jersey: Prentice Hall, 1960), 46.

754 Nasr, "Islamic Worldview and Modern Science," 22. Also, see Seyyed Hossein

References

List of works of Nasr

Nasr, Seyyed Hossein. "In the Beginning was Consciousness." In *The Essential Seyyed Hossein Nasr*, ed. William C. Chittick. Bloomington, IN: World Wisdom, 2007: 223-29.

______. "On Biological Origins." Interview by Muzaffar Iqbal, in *Islam, Science, Muslims, and Technology: Seyyed Hossein Nasr in Conversation with Muzaffar Iqbal.*

Alberta, Canada: Al-Qalam Publishing, 2007:149-174.

______. "Contemplation and Nature in the Perspective of Sufism." In *Islamic Life and Thought*. Chicago: ABC International Group, Inc., 2001: 200-06.

______. "Contemporary Man between the Rim and the Axis." In *Science and the Myth of Progress*, ed. Mehrdad M. Zandi. Bloomington, Indiana: World Wisdom, 2003:93-109.

______. "Cosmology." In *Different Aspects of Islamic Culture* (part 1), ed. A.Y.al-Hassan Paris: UNESCO Publishing, 2001: 361- 400.

______. "The Cosmos as Theophany." In *The Essential Seyyed Hossein Nasr*, ed. William C. Chittick. Bloomington, Indiana: World Wisdom, 2007: 189-202.

______. "The Ecological Problem in the Light of Sufism: The Conquest of Nature and the Teachings of Eastern Science." In *Sufi Essays*, 3rd edition. Chicago: ABC International Group, 1999: 152-63.

______. *The Encounter of Man and Nature: The Spiritual Crisis of Modern Man.* London: George Allen and Unwin, 1968.

______. "On the Environmental Crisis." Interview by Muzaffar Iqbal, in *Islam, Science, Muslims, and Technology: Seyyed Hossein Nasr in Conversation with Muzaffar Iqbal*. Alberta, Canada: Al-Qalam Publishing, 2007: 119-148.

______. *The Garden of Truth: The Vision and Promise of Sufism, Islam's Mystical Tradition*. New York: HarperCollins Publishers, 2007.

______. *The Heart of Islam: Enduring Values for Humanity*. New York: HarperCollins, 2002.

Nasr, "Traditional Man, Modern man, and the Environmental Crisis," *Sophia*, Vol 12, No. 2, Fall/Winter 2006, p. 33.

______. *Ideals and Realities of Islam,* 2nd edition. Chicago: Kazi Publications, 2000.

______. "An Intellectual Autobiography." In *The Philosophy of Seyyed Hossein Nasr,* The Library of Living Philosophers, Volume XXVII, eds. Lewis Edwin Hahn, Randall E. Auxier, Lucian W. Stone, Jr. Chicago and La Salle: Open Court Publishing Company, 2001: 3-85.

______. *An Introduction to Islamic Cosmological Doctrines: Conceptions of Nature and Methods Used for the Ikhwan al-Safa, Al-Biruni,and Ibn Sina.* Cambridge, Massachusetts: Harvard University Press, 1964.

______. "Islam, the Contemporary Islamic World, and the Environmental Crisis." In *Islam and Ecology: A Bestowed Trust,* eds. Richard Foltz, Frederick Danny and Azizan Baharuddin. Harvard University Press, 2003: 85-105.

______. "Islam and Modern Science." In *Islam and Contemporary Society,* ed. Salem Azzam London: Longman Group, 1982: 177-90.

______. "Islam, Muslims, and Modern Technology." Interview by Muzaffar Iqbal, *Islam & Science,* Vol. 3 (Winter 2005) No.2: 109-126.

______. *Islam and the Plight of Modern Man,* 2nd edition. Chicago: ABC International, 2001. 338

______. "Islam and the Problem of Modern Science." In *An Early Crescent: The Future Knowledge and Environment in Islam,* ed. Ziauddin Sardar. London: Mansell, 1998: 127-139.

______. "Islam, Science, and Muslims." Interview by Muzaffar Iqbal, in *Islam, Science, Muslims, and Technology: Seyyed Hossein Nasr in Conversation with Muzaffar Iqbal.* Alberta, Canada: Al-Qalam Publishing, 2007: 51-89.

______. *Islamic Art and Spirituality.* New York: State University of New York Press, 1987.

______. *Islamic Philosophy from its Origin to the Present: Philosophy in the Land of Prophecy.* Albany: State University of New York, 2006.

______. "Islamic Philosophy – Reorientation or Re-understanding." In *Islamic Life and Thought.* Chicago: ABC International Group, Inc., 2001: 153-57.

______. *Islamic Science: An Illustrated Study.* World of Islam Festival Publishing Company, 1976.

______. "The Islamic View of the Universe", in *Islamic Culture, vol. I, Foundation of Islam,* ed. Z. Ansari, Paris: UNESCO (in Press), manuscript: 1-41.

______. "Islamic worldview and Modern Science." *Islamic Thought and Scientific Creativity,* Vol. 7 (1), 1996: 7- 22.

______. *Knowledge and the Sacred.* Albany: State University of New York, 1989.

______. "The Life, Doctrines and Significance of Sadr al-Din Shirazi (Mulla Sadra)." In *Islamic Life and Thought.* Chicago: ABC International Group, Inc., 2001: 158-68.

______. *Man and Nature: The Spiritual Crisis in Modern Man*. Chicago: ABC International Group, Inc., 1997.

______. "Mulla _adra: His teachings." In *History of Islamic Philosophy, Part I,* eds. Seyyed Hossein Nasr and Oliver Leaman. London: Routledge, 1996: 643-661.

______. *The Need for a Sacred Science.* New York: State University of New York, 1993.

______. "Pertinence of Studying Islamic Philosophy Today." In *Islamic Life and Thought.* Chicago: ABC International Group, Inc., 2001: 145-52.

______. *The Pilgrimage of Life and the Wisdom of Rumi*. Oakton, VA: The Foundation of Traditional Studies, 2007.

______. "In the Quest of the Eternal Sophia." In *The Complete Bibliography of the Works of Seyyed Hossein Nasr,* ed. Mehdi Aminrazavi and Zailan Morris. Kuala Lumpur, 1994: xxvii-xxxii.

______. "The Question of Cosmogenesis – The Cosmos as a Subject of Scientific Study," *Islam and Science,* Vol. 4(Summer 2006) No.1: 43-59.

______. "Reflections on Methodology in the Islamic Sciences," *Hamdard Islamicus,* Vol. III No.3, (1980): 3-13.

______. "Religion and the Environmental Crisis." In *The Essential Seyyed Hossein Nasr,* ed. William C. Chittick. Bloomington, Indiana: World Wisdom, 2007: 29-39.

______. *Religion and the Order of Nature.* New York: Oxford University Press, 1996.

______. "Reply to Huston Smith." In *The Philosophy of Seyyed Hossein Nasr,* The Library of Living Philosophers, Volume XXVII, eds. Lewis Edwin Hahn, Randall E. Auxier, Lucian W. Stone, Jr. Chicago and La Salle: Open Court Publishing Company, 2001: 159-67.

______. "Reply to Sallie King." In *The Philosophy of Seyyed Hossein Nasr,* The Library of Living Philosophers, Volume XXVII, eds. Lewis Edwin Hahn, Randall E. Auxier, Lucian W. Stone, Jr. Chicago and La Salle: Open Court Publishing Company, 2001: 221-231.

______. *_adr al-Din Shirazi and his Transcendent Theosophy: Background, Life and Works*. Tehran, Iran: Imperial Iranian Academy of Philosophy, 1978.

______. *Science and Civilization in Islam,* 2nd edition. Cambridge, UK: The Islamic Texts Society, 2007.

______. *Sufi Essays*. 3rd edition. Chicago: ABC International Group, 1999.

______. "Theoretical Gnosis and Doctrinal Sufism and their Significance Today." *Transcendent Philosophy,* Vol. 6, Dec 2005: 1-35.

______. *Three Muslim Sages*. New York: Caravan Books, 1964.

______. "Traditional Cosmology and Modern Science: Interview with Seyyed Hossein Nasr. " *Parabola,* Vol. 8.4: 104-120.

______. *Traditional Islam in the Modern World*. New York: Kegan Paul International, 1994.

______. "Traditional Man, Modern man, and the Environmental Crisis." *Sophia,* Vol 12, No. 2, Fall/Winter 2006: 28-33.

______. "Traditionalist Approach to Religion." In *The Essential Seyyed Hossein Nasr,* ed. William C. Chittick. Bloomington, Indiana: World Wisdom, 2007: 21-27.

______. *A Young Muslim's Guide to the Modern World*. Chicago: Kazi Publications, 1994.

Nasr, Seyyed Vali Reza. "Mawdudi and the Jama'at-i Islami: The Origins, Theory and Practice of Islamic Revivalism." In *Pioneers of Islamic Revival,* ed. Ali Rahnema, London: Zed Books Ltd., 1994: 98-124.

List of works of other different scholars

'Abduh, Muhammad. *The Theology of Unity,* trans. by Ishaq Musa'ad and Kenneth Cragg.

Abraham, Jun. rev. of *Limits to Growth: 30-Year Update* by Donella Meadows, Jorgen Randers and Dennis Meadows, *Natural Resources Forum* 29(2005): 179-80.

Acar, Rahim. "Avicenna's position concerns the Basis of the Divine Creative

Action." *The Muslim World*, Vol. 94, January 2004: 65-79.

Acikgenc, Alparslan. "Holistic Approach to Scientific Traditions." *Islam and Science,* Vol. 1 (June 2003) No. 1: 99-114.

Adams, Charles C. *Islam and Modernism in Egypt*. London: Oxford University Press, 1933.

Afrasiabi, Kaveh L. "Toward an Islamic Ecotheology." In *Islam and Ecology: A Bestowed Trust,* eds. Richard C. Foltz, Frederick M. Denny, and Azizan Baharuddin, Cambridge, MA: Harvard University Press, 2003: 281-96.

Ahmad, Ali. "Nigeria." In *Environmentalism in the Muslim World,* ed. Richard C. Foltz. New York: Nova Science Publishers, 2005: 75-83.

Ahwal, Ayman. "Saving Rivers and Forests in Malaysia." In *Environmentalism in the Muslim World,* ed. Richard C. Foltz. New York: Nova Science Publishers, 2005: 113-131.

Al-Afghani, Jamal al-Din. "An Islamic Response to Imperialism." In *Islam in Transition: Muslim Perspectives,* eds. John J. Donohue and John L. Esposito. 2nd edition. New York: Oxford University Press, 2006.

Al-Attas, Syed Muhammad Naquib. *Islam and Secularism*. Kuala Lumpur: International Institute of Islamic Thought and Civilization, 1993.

Al-Azmeh, Aziz. *Islams and Modernities*. London: Verso, 1996.

Al-Banna, Hasan. *Selected Writings of Hasan al-Banna Shaheed, trans. S.A. Qureshi. Karachi:* International Islamic Publishers, 1983.

Al-Ghazzali, Abu Hamid. "The Elaboration of the Marvels of the Heart." In Knowledge of God in Classical Sufism, trans. John Renard. Classics of Western Spirituality. New York: Paulist Press, 2004: 298-326.

Al-Iskandari, Ibn 'Ata'illah. *The Book of Wisdom*, Chapter 1, verse 14, trans. Victor Danner. New York: Paulist Press, 1978.

Al-Kalabadhi, Abu Bakr. "The Exploration of Sufi Teachings (*Kitab at-ta'arruf*)." In *Knowledge of God in Classical Sufism: Foundations of Islamic Mystical Theology,* trans. John Renard. Classics of Western Spirituality. New York: Paulist Press, 2004: 100-111.

Al-Makki, Abu Talib. "The Sustenance of Hearts (*Qut al-qulub*)." In *Knowledge of God in Classical Sufism: Foundations of Islamic Mystical Theology,* trans. John Renard. Classics of Western Spirituality. New York: Paulist Press, 2004: 112-263.

Anderson, Fulton H. "Editor's Introduction," In *The New Organon and*

Related Writings, ed. Fulton H. Anderson, New Jersey: Prentice Hall, 1960: vii-xxxviii.

An-Nadwi, Mohammed Akram. *Madrasah Life*. London: Turath Publishing, 2007.

Ansari, 'Abd Allah. "The Hundred Fields: Resting Places of the Wayfarers (*Sad Maydan*)." In *Knowledge of God in Classical Sufism: Foundations of Islamic Mystical Theology,* trans. John Renard. Classics of Western Spirituality. New York: Paulist Press, 2004: 294-297.

Appleyard, Bryan. *Understanding The Present: An Alternative History of Science*. London: Tauris Parke Paperbacks, 2004.

Ardalan, Nader. and Bakhtiar, Laleh. *The Sense of Unity.* Chicago: The University of Chicago

Ayoub, Mahmoud M. *The Quran and Its Interpreters,* Vol. I. Albany, NY: State University of New York, 1984.

Azizan Baharuddin. Cambridge, MA: Harvard University Press, 2003: 499-510.

Bacon, Francis. "The New Organon." In *The New Organon and Related Writings,* ed. Fulton H. Anderson. New Jersey: Prentice Hall, 1960: 33-268.

Bacon, Francis. "The Great Instauration." In *The New Organon and Related Writings,* ed. Fulton H. Anderson. New Jersey: Prentice Hall, 1960: 3-29.

Bakar, Osman. "Civilizational Dialogue on Environmental Philosophy: Nasr's 'Man and Nature' Revisited." In *The Beacon of Knowledge: Essays in Honor of Seyyed Hossein Nasr,* ed. Mohammad H. Faghfoory. Louisville, Kentucky: Fons Vitae, 2003: 291-303.

Bakhtiar, Laleh. *Introduction to Al-Ghazzali on Trust and the Unity of God, trans. Muhammad Nur Abdus Salam.* Chicago: Great Books of the Islamic World, 2002: 3-13..

Behe, Michael J. *The Edge of Evolution: The Search for the Limits of Darwinism. New York: Free Press, 2007.*

Bennett, Jane. *Unthinking Faith and Enlightenment: Nature and the State in a Post-Hegelian Era.* New York: New York University Press, 1987.

Berkey, Jonathan P.*"Madrasas Medieval and Modern: Politics, Education, and the Problem of Muslim Identity. "In Schooling Islam: The Culture and Politics of Modern Muslim Education, eds. Robert W. Hefner and Muhammad Qasim Zaman.* Princeton: Princeton University Press, 2007: 40-60.

Berlinski, David. *The Devil's Delusion: Atheism and Its Scientific Pretensions.* New York: Crown Forum, 2008.

Berry, Wendel. *"Christianity and the Survival of Creation." In Seeing God Everywhere: Essays on Nature and the Sacred, ed. Barry McDonald.* Bloomington, IN: World Wisdom, 2003: 53-70.

Bestowed Trust. Cambridge: Harvard University Press, 2003.

Blackburn, Steven. *rev. of Islamic Philosophy from Its Origin to the Present: Philosophy in the Land of Prophecy by Seyyed Hossein Nasr,* Religion and Theology, 14:3 (2007): 436-40.

Blanchard, Elodie Vielle. *"Modelling the Future: An Overview of the 'Limits to Growth' Debate."* Centaurus 2010: 52: 91-116.

Buell, Frederick. *From Apocalypse to Way of Life.* New York and London: Routledge, 2003.

Burckhardt, Titus. *An Introduction to Sufi Doctrine, trans. D.M. Matheson.* Lahore: Sh. Muhammad Ashraf Publishers, 1996.

Carvalho IV, John J. "Overview of the Structure of a Scientific Worldview." *Zygon,* vol. 41, no. 1 (March 2006): 113-124.

Cassirer, Ernest. *The Individual and the Cosmos in Renaissance Philosophy,* trans. Mario Domandi. Chicago: University of Chicago, 1963.

Catherine, D., Nature, Theophany and the Rehabilitation of Consciousness, 2007, electronic version only, accessed on www.nuradeen.com/Contributions/NatureTheophany.htm Catherine, D. personal communication (email), March 2009.

Charles, Prince of Wales, 'A Sense of the Sacred: Building Bridges Between Islam and the West' in: Sophia, The Journal of Traditional Studies, Volume 3, Number 1; Summer 1997. This revised version of the text is printed in Sacred Web with the express permission of the author. See: http://www.sacredweb.com/online_articles/

Cheetham, Tom, Green Man, Earth Angel, The Prophetics Tradition and the Battle for the Soul of the World, New York: State University of New York Press, 2005.

Chenu, Marie-Dominique. *Nature, Man and Society in the Twelfth Century.* Cambridge, MA: Medieval Academy of America, 1997.

Chittick, W.C. (eds) The Essential Seyyed Hossein Nasr, The perennial Philosophy series, Bloomington: World Wisdom, 2007.

Chittick, William C. "Ibn 'Arabi and His School." In *Islamic Spirituality: Manifestations*, ed. Seyyed Hossein Nasr. New York: The Crossroad Publishing Company, 1997: 49-79.

Christian and Islamic Middle Ages. Oxford: University of Oxford, 2004.

Christians, Clifford G. "Jacques Ellul's Conversions and Protestant Theology." Journal of Media and Religion, 5(3): 147-60.

Commins, David. "Hasan al-Banna(1906-1949)." In Pioneers of Islamic Revival, ed. Ali Rahnema. London: Zed Books, 1994: 125-53.

Cook, Bradley J. "Islamic Versus Western Conception of Education: Reflections on Egypt." International Review of Education, 45(3/4): 1999: 339-357.

Corbin, Henry. Avicenna and the Visionary Recital, trans. Willard R. Trask. New York: Pantheon Books Inc., 1960.

Cornwall, Andrea. "Buzzwords and fuzzwords: deconstructing development discourse." *Development in Practice*, Volume 17, Numbers 4-5, August 2007: 471-484.

Critchlow, Keith. "Keyonte Speech." In *The Beacon of Knowledge: Essays in Honor of Seyyed Hossein Nasr*. ed. Mohammad H. Faghfoory. Louisville, Kentucky: Fons Vitae, 2003: xlvii-liii.

Dagli, Caner. "Mulla Sadra's Epistemology and the Philosophy of Physics." In Perception According to Mulla Sadra, ed. Seyed G. Safavi. London: Salman-Azadeh Publication, 2002: 245-258.

Dakake, Maria Massi. "The Soul as Barzakh: Substantial Motion and Mulla Sadra's Theory of Human Becoming." The Muslim World, Vol. 94, January 2004: 107-30.

Dangor, Suleman. "Islamization of Disciplines: Towards an indigenous educational system." Educational Philosophy and Theory, Vol. 37, No.4, 2005: 519-531.

De Jong, Frederick. and Radtke, Bernd. Introduction to Islamic Mysticism Contested: Thirteen Centuries of Controversies and Polemics, eds. Frederick De Jong and Bernd Radtke. Leiden: Brill, 1999: 1-21.

Dear, Peter. Revolutionizing the Sciences: European Knowledge and Its Ambitions, 1500 1700. 2nd edition. Princeton: Princeton University Press, 2009.

Dewey, John. "The Influence of Darwinism on Philosophy." In The Essential Dewey: Pragmatism, Education, Democracy Vol. 1(Bloomington, IN: Indiana University Press, 1998), 39-45.

Dewitt, Richard. Worldviews: An Introduction to the History and Philosophy of Science. Oxford, UK: Blackwell, 2004.

Doran, Peter. and Zimmerman, Maggie K. "Examining the Scientific Consensus on Climate Change." EOS American Geophysical Union, Vol. 90 No.3 20 January 2009: 22-23.

Dupre, Louis. "Spiritual Life and the Survival of Christianity." Cross Currents; Fall 1998, Vol. 48 issue 3: 381-91.

Eberley, John. Al-Kimia: The mystical Islamic Essence of the Sacred Art of Alchemy. Hillsdale, NY: Sophia Perennis, 2004.

Edis, T., An Illusion of Harmony, Science and Religion in Islam, New York: Prometheus Books, 2007.

Edson, E. and Savage-Smith, E. Medieval Views of the Cosmos: Picturing the Universe in the

El-Ansary, Walid. "The Traditionalist Critique of Industrial Capitalism." *Sophia* Vol 12, no. 1.

Eliade, Mircea. The Sacred and the Profane: The Nature of Religion, trans. Willard Trask. New

Ellul, Jacques. The Technological Society. New York: Random House, 1964.

Ernst, Carl. Rethinking Islam in the Contemporary World. Edinburgh: Edinburgh University Press, 2004.

Ezzat, Heba Raouf. "Secularism, the State and the Social Bond: The Withering Away of the Family." In Islam and Secularism, eds. John Esposito and Azzam Tamimi. London: C. Hurst & Co., 2000: 124-137.

Fadiman, James. and Frager, Robert, ed. Essential Sufism. New York: HarperCollins Publishers, 1997.

Faghfoory, Mohammad H. Introduction to The Beacon of Knowledge: Essays in Honor of Seyyed Hossein Nasr, ed. Mohammad H. Faghfoory. Louisville, Kentucky: Fons Vitae, 2003: xxvii-xl.

Fathy, Hasan. Architecture for the Poor. Chicago: University of Chicago Press, 2000.

Feder, Toni. "International Research University Open in Saudi Arabia." Physics Today, November 2009: 24-25.

Feyerabend, Paul. Against Method. 4th edition. London: Verso, 2010.

Foltz, R.C. Denny, F. Baharuddin, A. (eds), Islam and Ecology: A Bestowed Trust, Cambridge, Massachusetts: Harvard University Press, 2003.

Foltz, R.C., Environmentalism in the Muslim World, New York: Nova Science Publishers, 2005.

Foltz, R.C., personal communication (email), 17 September, 2 October, 4 December 2008.

Foltz, Richard C. Introduction to Islam and Ecology: A Bestowed Trust. eds. Richard C. Foltz,

Foltz, Richard C., Denny, Frederick M., and Baharuddin, Azizan, eds. Islam and Ecology: A

Foltz, Richard C., ed. Environmentalism in the Muslim World. New York: Nova Science Publishers, Inc., 2005.

Frangsmyr, Tore., Heilborn, J.L., Rider, Robin E. eds. The Quantifying Spirit in the 18th Century. Berkely: University of California, 1990.

Frederick M. Denny, and Azizan Baharuddin. Cambridge, MA: Harvard University Press, 2003: xxxvii-xliii.

Fuller, Graham E., The Future of Political Islam, New York: Palgrave MacMillian, 2003.

Gabrieau, Marc. "Criticizing the Sufis: The Debate in Early-Nineteenth-Century India." In Islamic Mysticism Contested: Thirteen Centuries of Controversies and Polemics, eds. Frederick De Jong & Bernd Radtke. Leiden: Brill, 1999: 452-467.

Gardener, Gary. Inspiring Progress: Religious Contributions to Sustainable Development. New York: W.W. Norton, 2006.

Gesink, Indira Falk. "Islamic Reformation: A History of Madrasa Reform and Legal Change in Egypt." In Islam and Education: Myths and Truth, eds. Wadad Kadi and Victor Billeh. Chicago: University of Chicago, 2007: 19-39.

Godfrey-Smith, Peter. *Theory and Reality: An Introduction to the Philosophy of Science.* Chicago: The University of Chicago, 2003.

Goldsmith, E., 'Religion in the Millenium' in: The Ecologist, special issue on Cosmic Religion, November 1999.

Goldstone, Jack. *Why Europe?* Boston: McGraw Hill Higher Education, 2008.

Golshani, Mehdi. "Does Science Offer Evidence of a Transcendent Reality?" Islam and Science, Vol. 1 (June 2003) No. 1: 45-58.

Graham, W., 'Traditionalism in Islam, an essay in interpretation' in: Journal of Interdisciplinary History, vol.23, no.3, p495-522, 1993.

Grant, Edward. Science and Religion: 400 BC – AD 1550. Baltimore, MD: The Johns Hopkins University Press, 2004.

Grant, George. Technology and Justice. Concord, Ontario: Anansi Press, 1986.

Guénon, René. The Crisis of the Modern World, trans. Marco Pallis, Arthur Osborne and Richard C. Nicholson. Hillsdale, NY: Sophia Perennis, 2001.

Gutas, D., 'The Study of Arabic Philosophy in the Twentieth Century, an essay on the historiography of Arabic Philosophy', in: British Journal of Middle Eastern Studies, 2002, p. 29(1), 5-25.

Gutas, Dimitri. "Islam and Science: A False Statement of the Problem." Islam & Science, Vol. 1 (December 2003) No. 2: 215-220.

Haddad, Yovonne. "Muhammad 'Abduh: Pioneers of Islamic Reform." In Pioneers of Islamic Revival, ed. Ali Rahnema. London: Zed Books Ltd., 1994.

Hafez, M.H. "Explaining the Origins of Islamic Resurgence: Islamic Revivalism in Egypt and Indonesia." The Journal of Social, Political, and Economic Studies; Fall 1997: 295-324.

Hahn, L.E., Auxier, R.E., Stone., L.W Jr (eds), The Philosophy of Seyyed Hossein Nasr, The Library of Living Philosophers, Volume XXVII, Chicago and La Salle: Open Court Publishing Company, 2001.

Haleem, Harfiyah Abdel. Islam and the Environment. London: Ta-Ha Publishers Ltd., 1998.

Halstead, J. Mark. "An Islamic concept of education," in Comparative Education, Vol. 40, No. 4, November 2004: 517-529.

Halteman, James. rev. of The Bridge at the Edge of the World by James Gustave Speth, Christian Century, December 16, 2008: 38-9.

Hanegraaff, W. New Age Religions and Western Culture – Esotericism in the Mirror of secular Thought, Leiden: Brill, 1996. p58-59.

Hanegraaff, W.J., eds. Dictionary of Gnosis &Western Esotericism, Leiden: Brill, 2006.

Haq, S. Nomanul. Names, Natures and Things: The Alchemist and his Kitab al-Ahjar (Book of Stones). Boston: Kluwer Academic Publishers, 1994.

Harding, Brian. "Scepticism, Illumination and Christianity: In Augustine's Contra Academicos," Augustinian Studies 34:2 (2003): 197-212.

Harrison, Peter. "Subduing the Earth: Genesis 1, Early Modern Science, and the Exploitation of Nature" The Journal of Religion, 1999: 86-109;

Have, A.M.J. ten (eds), Environmental Ethics and International Policy (2006), Ethics series, UNESCO publishing, Paris, 2006.

Heath, Peter. Allegory and Philosophy in Avicenna. Philadelphia: University of Pennsylvania Press, 1992.

Heilbron, John L. "Coming to Terms with the Scientific Tradition." European Review, Vol. 15, No. 4, 2007: 473-89.

Henry, John. The Scientific Revolution and the Origins of Modern Science. Basingstoke, UK: Macmillan, 1997.

Hermansen, M, personal communication (email), 17, 18, 21 December 2008.

Hermansen, M., 'The Academic Study of Sufism at American Universities', American Journal of Islamic Social Sciences, Vol.24, 3, 2007.

Hillman, Mayer. with Fawcett, Tina. and Rajan, Sudhir Chella. How Can We Save the Planet. New York: St. Martin's Press, 2008.

Hodgson, Marshall G. S. The Venture of Islam: Conscience and History in a World Civilization, Vol. 2. Chicago: The University of Chicago Press, 1977.

Hoinacki, Lee. "The Trajectory of Ivan Illich." Technology & Society, Vol. 23, No.5, October 2003: 382-89.

Holmyard, E.J. Alchemy. New York: Dover Publications, 1990.

Hoodbhoy, Pervez. Islam and Science: Religious Orthodoxy and the Battle for Rationality. London: Zed Books Ltd., 1991.

Hourani, Albert H. Arabic Thought in the Liberal Age 1798-1839. Cambridge: Cambridge University Press, 1983.

Huxley, Aldous, The Perennial Philosophy, London: Waldo & Windus, 1980.

Ibn Sina, Abu 'Ali al-Hussain. Remarks and Admonitions, Part Four. In Ibn Sina and Mysticism: Remarks and Admonitions, Part Four, trans. Shams Inati. London: Kegan Paul

Illich, Ivan. Tools for Conviviality. New York, London: Harper & Row, 1973.

International, 1996.

Iqbal, M., 'The Islamic Perspective on the Environmental Crisis: Seyyed Hossein Nasr in Conversation with Muzaffar Iqbal' in: Islam & Science, Vol.5 Issue 1 p 75(22), June 22, 2007, Thomson Gale (accessed:www.thefreelibrary.com)

Iqbal, Muzaffar. Islam and Science. Hampshire and Burlington: Ashgate, 2002.

Isenberg, Shaya., and Thursby, Gene. "Perennial Philosophy," in *The Encyclopedia of Religion and Nature*, ed. Bron Taylor. New York: Thoemmes Continuum, 2005: 1269-72.

Izzi Dien, Mawil. *The Environmental Dimension of Islam*. Cambridge, England: The Lutterworth Press, 2000.

Jacob, Margaret. *Scientific Culture and The Making of the Industrial West*. New York and Oxford: Oxford University Press, 1997.

James, Christopher. and Fitzerald, Joseph, the 5th Lord Northbourne, eds. *Of the Land and the Spirit: The Essential Lord Northbourne on Ecology*. Bloomington, IN: World Wisdom, 2008.

Jawad, Haifaa "Seyyed Hossein Nasr and The Study of Religion in Contemporary Society." *The American Journal of Islamic Social Sciences*, 22:2, 2005: 49-68.

Jenkins, Willis. "After Lynn White: Religious Ethics and Environmental Problems." *Journal of Religious Ethics*, 37.2, 2009: 283-309.

Johansen, Julian. *Sufism and Islamic Reform in Egypt: The Battle for Islamic Tradition*. Oxford: Clarendon Press, 1996.

John Renard. Classics of Western Spirituality. New York: Paulist Press, 2004: 264-285.

Johnston, Douglas., Hussain, Azhar. and Catildi, Rebecca. *Madrasa Enhancement and Global Security*. International Center for Religion and Diplomacy, 2008.

Jong F.de, Radtke, B. Islamic Mysticism Contested: Thirteen Centuries of Controversies and Polemics, ed. Leiden, Brill, 1999.

Kalin, Ibrahim. "Between Physics and Metaphysics," in *Islam & Science*, Vol. 1 (2003) No.1: 59-90.

Kamali, Mohammad Hashim. Principles of Islamic Jurisprudence. Cambridge: Islamic Texts Society, 2008.

Kateb, George. "Locke and the Political Origins of Secularism," Social Research, Vol. 76: No 4: Winter 2009: 1001-1034.

Keddie, Nikki R. An Islamic Response to Imperialism. Berkeley: University of California Press, 1983.

Khalid, Fazlun. "Applying Islamic Environmental Ethics." In Environmentalism in the Muslim World. ed. Richard C. Foltz. New York: Nova Science Publishers, 2005: 87-111.

Khan, Muhammad Salim. Islamic Medicine. London: Routledge, 2008.

Knysh, Alexander. "Contextualizing the Salafi-Sufi Conflict." Middle Eastern Studies, Vol. 43, No. 4, July 2007: 503-530.

Kuhn, Thomas S. The Structure of Scientific Revolutions, 3rd edition. Chicago: The University of Chicago Press, 1996.

Lapidus, Ira M. A History of Islamic Societies, 12th edition. Cambridge: University Press, 1997.

Laude, Patrick. "Seyyed Hossein Nasr in the Context of the Perennialist School." In Beacon of Knowledge: Essays in Honor of Seyyed Hossein Nasr, ed. Mohammad Faghfoory. Louisville, KY: Fons Vitae, 2003: 245-259.

Leaman, Oliver. "In search of Tradition: Islamic Art and Science in the thought of Seyyed Hossein Nasr." In *Beacon of Knowledge: Essays in Honor of Seyyed Hossein Nasr*. ed. Mohammad Faghfoory. Louisville, KY: Fons Vitae, 2003: 305-315.

Lekkas, Georgios. "Plotinus: Towards an Ontology of Likeness (On the One and the Nous)." *International Journal of Philosophical Studies*, vol. 13(1): 53-68.

Leopold, Aldo. *A Sand County Almanac*. New York: Ballantine Books, 1970.

Lindberg, David C. *The Beginnings of Western Science: The European Scientific Tradition in Philosophical, Religious, and Institutional Context, Prehistory to A.D. 1450*, 2nd edition. Chicago: The University of Chicago Press, 2007.

Lings, M., A Sufi Saint of the Twentieth Century: sheikh Ahmad al-Alawi, Berkley and Los Angeles: University of California Press, 1961, 1974.

Lings, Martin. *Ancient Beliefs and Modern Superstitions,* 2nd edition. London: Unwin Paperbacks, 1980.

Llewellyn, Othman Abd-ar-Rahman. "The Basis for a discipline of Islamic Environmental Law." In *Islam and Ecology: A Bestowed Trust*, eds. Richard C. Foltz, Frederick M. Denny, and Azizan Baharuddin. Cambridge, MA: Harvard University Press, 2003: 185-247.

Locke, John. "Of Innate Notions." In *The Age of Enlightenment: The 18th Century Philosophers*. ed. Sir Isaiah Berlin. New York: The New American Library, 1956: 34-40.

London: George Allen and Unwin Ltd., 1966.

Lory, P., Lecture held at Leiden University, Honours Class, 11 April 2008 and email correspondence, 30 May 2008.

Louth, Andrew. *Maximus the Confessor*. New York: Routledge, 1996.

Lumbard, Joseph E. B. "The Decline of Knowledge and the Rise of Ideology in the Modern Islamic World." In *Islam, Fundamentalism and Betrayal of Tradition*. ed. Joseph E.B. Lumbard, Bloomington, IN: World Wisdom, 2004: 39-77.

Malik, J, Hinnells, J, Sufism in the West, London and New York: Routledge, 2006.

Manzoor, S. Parvez. "Desacralising Secularism." In *Islam and Secularism in the Middle East,* eds. John Esposito and Azzam Tamimi, London: C. Hurst & Co., 2000: 81-96.

Martines, Lauro. "The Protean Face of Renaissance Humanism." *Modern Language Quarterly*, 1990; 51: 105 -121

Masood, Ehsan. "An Islamist Revolution." *Nature* Vol. 444, 2 Nov 2006: 22-25.

Mattson, Ingrid. *The Story of the Quran: Its History and Place in Muslim Life.* Malden, MA: Blackwell Publishing, 2008.

Mawdudi, Abu'l 'Ala'. *West versus Islam,* trans. Waqar Ahmad Gardezi & Abdul Waheed Khan. Markazi Maktaba Islami Publishers, 2000.

McCloskey, Deirdre N. *Bourgeois Dignity: Why Economics Can't Explain the Modern World*. Chicago and London: University of Chicago Press, 2010.

McDonald, Barry, ed. *Seeing God Everywhere: Essays on Nature and the Sacred*. Bloomington, IN: World Wisdom, 2003.

Meadows, Donella H., Meadows, Dennis L., Randers, Jorgen. and Behrens III, William W. *The Limits to Growth*. New York: Universe Books, 1972.

Meadows, Donella H., Randers, Jorgen. and Meadows, Dennis. *Limits to Growth: The 30-Year Update.* White River Junction, Vermont: Chelsea Green, 2004.

MeClellan III, James E. and Dorn, Harold. *Science and Technology in World History.* Baltimore: The Johns Hopkins University Press, 2006.

Meyer, Stephen. *Signature in the Cell: DNA and the evidence for Intelligent Design*. New York: HarperCollins, 2009.

Milton-Edwards, Beverley. *Islamic Fundamentalism Since 1945*. New York: Routledge, 2005.

Minteer, Ben A. and Collins., James P. "From Environmental to Ecological Ethics: Toward a Practical Ethics for Ecologists and Conservationists." *Sci Eng Ethics*(2008):483-501.

Mitchell, Richard. *The Society of Muslim Brothers*. Oxford: Oxford University Press, 1993.

Moaddel, Monsoor. *Islamic Modernism, Nationalism and Fundamentalism.* Chicago: University of Chicago Press, 2005.

Mokyr, Joel. *The Enlightened Economy: An Economic History of Britain 1700-1850*. New Haven: Yale University Press, 2009.

Monastra, Giovanni. "Seyyed Hossein Nasr: Religion, Nature, and Science." In *The Philosophy of Seyyed Hossein Nasr,* The Library of Living Philosophers, Volume XXVII, eds. Lewis Edwin Hahn, Randall E. Auxier, Lucian W. Stone, Jr. Chicago: Open Court Publishing Company, 2001: 493-509.

Moran, Bruce T. *Distilling Knowledge: Alchemy, Chemistry, and the Scientific Revolution*. Cambridge, MA: Harvard University Press, 2005.

Moris, Zailan. *Revelation, Intellectual Intuition and Reason in the Philosophy of Mulla Sadra: An Analysis of al-Hikmah al-'Arshiyyah*. Richmond: Curzon, 2003.

Morris, James Winston. *The Reflective Heart*. Louisville, KY: Fons Vitae, 2005.

Morris, James. "Ibn Arabi and his Interpreters." Part II-A, *Journal of the American Oriental Society*, 106 (1986): 733-56.

Mousalli, Ahmad S. *Radical Islamic Fundamentalism*. Beirut: American University of Beirut, 1995.

Murata, Sachiko. and Chittick, William C. *The Vision of Islam*. St. Paul, Minnesota: Paragon House, 1994.

Murata, Sachiko. and Chittick, William C. *The Vision of Islam*. St. Paul, Minnesota: Paragon House, 1994.

Naeem, Fuad S. " A Traditional Islamic Response to the Rise of Modernism," In *Islam, Fundamentalism and Betrayal of Tradition,* ed. Joseph E.B. Lumbard. Bloomington, IN: World Wisdom, 2004: 79-116.

Needleman, Jacob. *A Sense of The Cosmos: Scientific Knowledge and Spiritual Truth*. Rhinebeck, New York: Monkish Book Publishing, 2003.

Neihardt, John, ed. *Black Elk Speaks*. London: Abacus, 1974.

Nielsen, Torben. "The State, the Market and the Individual." *Acta Sociologica* 1986(29): 283 302.

Northbourne, Lord. "Religion and Tradition." In *The Underlying Religion: An Introduction to the Perennial Philosophy,* eds. Martin Lings and Clinton Minnaar. Bloomington, IN: World Wisdom, 2007: 3-10.

Norton, Bryan. "Beyond Positivist Ecology: Towards an Integrated Ecological Ethics." *Sci Eng Ethics* (2008) 14: 581-592.

Oelschlaeger, Max. *Caring for Creation: An Ecumenical Approach to the Environmental Crisis.* New Haven: Yale University Press, 1994.

Oldmeadow, K., Traditionalism: Religion in the Light of the Perennial Philosophy, Shrilanka Institute of Traditional Studies, 2000.

Oldmeadow, Kenneth. *Traditionalism: Religion in the Light of the Perennial Philosophy*. Sri Lanka Institute of Traditional Studies, 2000.

Olson, Richard G. *Science and Scientism in Nineteenth-Century Europe.* Urbana, IL: University of Illinois, 2008.

Oreskes, Naomi. and Conway, Erik. *Merchants of Doubt: How a Handful of Scientists Obscured the Truth on Issues from Tobacco Smoke to Global Warming*. London: Bloomsbury Press, 2010. In *Climate Change: What it Means for Us, Our Children, and Our Grandchildren,* eds. Joseph Dimento and Pamela Doughman. Cambridge, MA: The MIT Press, 2007: 65-99.

Oviedo, Lluis. "Whom to blame for the Charge of Secularization." *Zygon,* vol. 40, no.2 (June 2005): 351-361.

Ozdemir, Ibrahim. "Turkey," In *Environmentalism in the Muslim World,* ed. Richard C. Foltz. New York: Nova Science Publishers, Inc., 2005: 17-36.

Peskes, Esther. "The Wahhabi yya and Sufism in the Eighteenth Century," Frederick De Jong & Bernd Radtke, eds. Islamic Mysticism Contested: Thirteen Centuries of Controversies and Polemics. Leiden: Brill, 1999: 145-161.

Petruccioli, Attilio. "Nature in Islamic Urbanism: The Garden in Practice and in Metaphor." In Islam and Ecology: A Bestowed Trust, eds. Richard C. Foltz, Frederick M. Denny, and

Pittenger, W. Norman. "The Christian Philosophy of John Scotus Erigena," The Journal of Religion, vol. 24, No. 4 (Oct., 1944): 246-57. Plato, Timaeus 3:28.

Pormann, Peter. and Savage-Smith, Emille. Medieval Islamic Medicine. Edinburgh: Edinburgh University Press, 2007.

Porter, Roy. "The Enlightenment in England." In The Enlightenment in National Context, eds. Roy Porter and Mikulas Teich. Cambridge: Cambridge University Press, 1981: 1-18.

Postman, Neil. Building a Bridge to the 18th Century: How the Past Can Improve Our Future. New York: Vintage Books, 1999.

Press, 1973. Aristotle, *De Anima*, Part II, 1.

Randall E. Auxier, Lucian W. Stone, Jr. Chicago: Open Court Publishing Company, 2001: 277-303.

Rees, Graham. and Jardine, Lisa, eds. Philosophical Studies, C.1611-C.1619, Vol. 6. New York: Oxford University Press, 1996.

Renard, John. Introduction to Knowledge of God in Classical Sufism: Foundations of Islamic Mystical Theology. Classics of Western Spirituality. New York: Paulist Press, 2004: 11-63.

Renard. Classics of Western Spirituality. New York: Paulist Press, 2004: 286-293.

Robinson, Francis. "Secularization, Weber and Islam." In Max Weber and Islam, eds. Toby E. Huff and Wolfgang Schluchter. New Brunswick, ME: Transaction Publishers, 1999: 231-245.

Rockstrom, Johan. et al, "A Safe Operating Space for Humanity", Nature, Vol 461 24 September 2009: 472-475.

Roper, Duncan. "The earth as a 'garden' for all creatures: Lynn White forty years on." Stimulus Vol 15 No 4 November, 2007:12-20. Rosenberg, Alex. Philosophy of Science: A Contemporary Introduction, 2nd edition. New York: Routledge, 2005.

Rosenthal, Franz. Science and Medicine in Islam: A concise Collection of Essays. Brookfield, VT: Ashgate Publishing Ltd., 1998.

Rossi, Vincent. "Sacred Cosmology in the Christian Tradition." The Ecologist Vol. 30; Issue 1, January 2000: 1-6.

Roszak, Theodore. Where the Wasteland Ends: Politics and Transcendence in Postindustrial Society. Berkeley, CA: Celestial Arts, 1989.

Rout, P. Francis and Bonaventure. Glasgow, UK: Fount Paperbacks, 1996.

Rustom, Mohammed. "Approaching Mulla adra as Scriptural Exegate: A Survey of Scholarship on His Qur'anic Works." Comparative Islamic Studies 4.1-2 (2008): 75-96.

Sachs, Wolfgang. Planet Dialectics: Explorations in Environment and Development. Halifax, Nova Scotia: Fernwood Publishing, 2001.

Salam, Abdus. "Islam and Science." In Ideals and Realities: Selected Essays of Abdus Salam, ed. C.H. Lai. Singapore: World Scientific; 2nd edition. 1987: 179-213.

Saliba, George. Islamic Science and the Making of the European Renaissance. Cambridge, MA: The MIT Press, 2007.

Sands, Kristin Zahra. Sufi Commentaries on the Qur'an in Classical Islam. New York: Routledge, 2006.

Sardar, Ziauddin. "Beyond the troubled relationship." Nature, Vol 448, 12 July 2007: 131-133.

Schimmel, A.M., Mystical dimensions of Islam, Chapel Hill: The University of North Carolina Press, 1975.

Schimmel, Annemarie As Through a Veil: Mystical Poetry in Islam. Oxford, UK: Oneworld, 2001.

Schliesser, Eric. "Wonder in the Face of Scientific Revolutions: Adam Smith on Newton's 'Proof' of Copernicanism." British Journal for the History of Philosophy 13(4) 2005: 697-732.

Schmidt, Lawrence. and Marratto, Scott. The End of Ethics in a Technological Society. Montreal & Kingston: McGill-Queen's University Press, 2008.

Schmidt-Biggeman, W., Philosophia Perennis: Historical Outlines of Western Spirituality in Ancient, Medieval and Early Modern Thought, International Archives of the History of Ideas, Dordrecht: Springer, 2004

Schopper, Herwig. "Where are the New Patrons of Science." Nature Vol. 444, 2 Nov 2006: 35 36.

Schouten, M.G.C. Spiegel van de Natuur: het natuurbeeld in cultuurhistorisch perspectief, Utrecht: KNNV Uitgeverij/ Staatsbosbeheer, 2005.

Schumacher, E. F. A Guide for the Perplexed. New York: HarperCollins, 2004.

Schumacher, E.F. A Guide for the Perplexed, London: Vintage, 1995.

Schumacher, E.F., Small is Beautiful, A Study of Economics as if People Mattered, London: Vintage, 1993.

Schuon, Frithjof. Logic and Transcendence, trans. Mark Perry, Jean-Pierre Lafouge, and James Cutsinger, ed. James Cutsinger. Bloomington, IN: World Wisdom, 2009.

Sedgwick, M, Against the Modern World: Traditionalism and the Secret Intellectual History of the Twentieth century, New York: Oxford University Press, 2004.

Sedgwick, M. personal communication (email), 19 June, 12, 16, 18 September, 10 October 2008.

Setia, A., 'The Inner Dimension of going Green: articulating an Islamic deep-ecology' in: Islam & Science, vol 5, no 2, winter 2007. This is an amended and extended version of a paper originally presented at the

International Conference on the role of Islamic States in a Globalized World, July 2007 by the Institute for Islamic Understanding (IKIM), Kuala Lumpur, Malaysia.

Setia, A., 'Three meanings of Islamic Science: toward Operationalizing Islamization of Science' in: Islam & Science, June 22, 2007

Shaikh, S., Kugle, S., 'To Love Every Life as Your Own: An Introduction to Engaged Sufism' in: Journal for Islamic Studies, Thematic Issue: Engaged Sufism, Vol.26, 2006.

Shepard, W., 'Islam and Ideology: Towards a Typology', in: International Journal of Middle Eastern Studies, p 307-336, Cambridge University Press, 1987

Spencer Trimingham, J. The Sufi Orders in Islam, New York: Oxford University Press, 1971.

Spring/Summer, 2006: 55-87.

Stenberg, Leif, The Islamization of Science: Four Muslim Positions, Developing an Islamic Modernity, New York, Coronet Books, 1996.

Stoddard, W., personal communication (email), 28 October, 6 November, 8 December 2008.

Stoddard, W., Remembering in a World of Forgetting: Thoughts on Tradition and Postmodernism, Bloomington, Indiana: World Wisdom, 2008.

Taylor, B. (eds), Encyclopedia of Religion and Nature, 2 volumes, New York: Continuum, 2005, 2008.

Temenos, A Sacred Trust – Ecology and Spiritual Vision, Temenos Academy, 2002.

Weinberg, Steven. Cosmology. Oxford and New York: Oxford University Press, 2008.

White, Jr., Lynn. "The historical roots of Our Ecological Crisis." Science Vol 155 (10 March 1967): 1203-1207.

White, Lynn, 'The Historical Roots of Our Ecological Crisis' in: Science 155:3767, 10 March 1967, p 1203-1207.

Wolf-Gazo, Ernest. "Nasr and the Quest for the sacred." In The Philosophy of Seyyed Hossein Nasr, The Library of Living Philosophers, Volume XXVII, eds. Lewis Edwin Hahn,

Wood, Barbara. E.F. Schumacher: his life and thought, New York: Harper & Row, 1984.

York: Harcourt, 1987.

Zaman, Muammad Qasim. The Ulama in Contemporary Islam. Princeton: Princeton University Press, 2002.

Zayd, Nasr, Abu, 'Trial of Thought: Modern Inquisition in Egypt, A Case Study', in: Drees, W.B., Koningsveld, P.S. Van (ed.), The Study of Religion and the Training of Muslim Clergy in Europe: Academic and Religious Freedom in the 21st century, Leiden: Leiden University Press, 2008.

Zeghal, Malika. "The 'Recentering' of Religious Knowledge and Discourse: The Case of al Azhar in the Twentieth-Century Egypt." In Schooling Islam: The Culture and Politics of Modern Muslim Education, eds. Robert W. Hefner and Muammad Qasim Zaman. Princeton: Princeton University Press, 2007: 107-130.

Zimmerman, John. "Roots of Conflict: The Islamist Critique of Western Value." The Journal of Social, Political and Economic Studies, Vol. 30, Number 4, 2005 : 425-58.

Public Lectures, Class Notes, and Conversations

Nasr, Seyyed Hossein. "Islam and Ecology." Lecture delivered at the Yale University, 9 April 2011.

Nasr, Seyyed Hossein. "Islam and the Environment." Lecture delivered at the Georgetown University School of Foreign Service, Qatar, 26 January 2009.

______. "Reading the Cosmic Qur'an." Lecture delivered to the Muslim Students' Association (MSA) at the George Mason University (USA), 27 March 2008.

Class notes for a graduate level course titled Man and Nature taught by S. H. Nasr at the George Washington University (USA); Spring Semester, 2004.

Conversation with Seyyed Hossein Nasr, The George Washington University (USA), 13 August 2009.

Conversation with Seyyed Hossein Nasr, Dedeman Hotel, Konya, Turkey, 17 May 2010.

Documents

C02 Emission From Fuel Combustion report of International Energy Agency(IEA), 2009 edition. Available from http://www.iea.org/co2highlights/co2highlights

The Earth Charter, available from: www.EarthCharter.org;

Al Hima: A Way of Life, available from http://cmsdata.iucn.org/downloads/al_Hima

The Key Scientific Developments Since The IPCC Fourth Assessment Report of June 2009, Intergovernmental Panel on Climate Change (IPCC) in *Science Brief* 2, June 2009, p. 5; available from www.pewclimate.org

Most relevant websites

Beacon of Knowledge Conference: http://www.beaconofknowledge.com/

Eye of the Heart: www.latrobe.edu.au/eyeoftheheart/

Foundation for Traditional Studies: www.traditional-studies.org

George Washington University, department of Religion: http://www.gwu.edu/~religion/faculty/Islamica: www.islamicamagazine.com

Muslim heritage: www.muslimheritage.com.

Muslim Philosophy: www.muslimphilosophy.com

Nasr Foundation: www.nasrfoundation.org

Sacred Web: www.sacredweb.com

Science and Religion in Islam: http://science-islam.net/

Studies in Comparative religion: www.studiesincomparativereligion.com.

Temenos Academy website: http://www.temenosacademy.org/

The Centre for Islam and Science: http://www.cis-ca.org/

Traditionalists.org: www1.aucegypt.edu/faculty/sedgwick/Trad/

Wisdom Books website: www.worldwisdom.com/

www.ingramcontent.com/pod-product-compliance
Lightning Source LLC
LaVergne TN
LVHW101505200926
848483LV00001B/229
9781804412947